Superstrings: A Theory of Everything?

Superstrings

A Theory of Everything?

Edited by

P.C.W. DAVIES

Professor of Theoretical Physics, University of Newcastle upon Tyne

JULIAN BROWN

Radio Producer in the BBC Science Unit, London

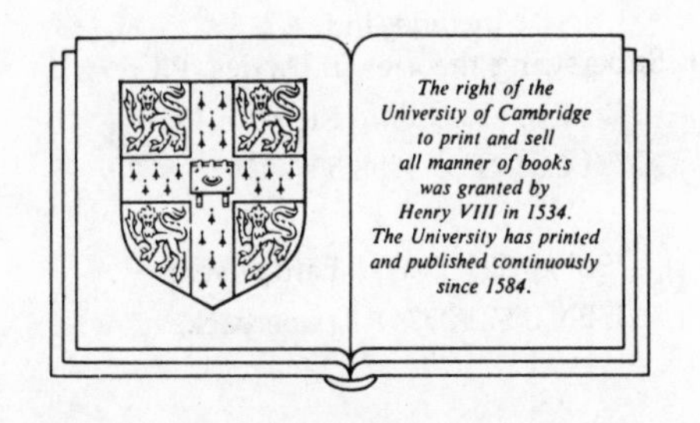

Cambridge University Press

Cambridge

New York New Rochelle

Melbourne Sydney

Published by the Press Syndicate of the University of Cambridge
The Pitt Building, Trumpington Street, Cambridge CB2 1RP
32 East 57th Street, New York, NY 10022, USA
10 Stamford Road, Oakleigh, Melbourne 3166, Australia

First published 1988
Reprinted 1988

Printed in Great Britain at the University Press, Cambridge

British Library Cataloguing in Publication data
Superstrings.
1. Elementary particles. Interactions. Superstring. Theories
I. Davies, P.C.W. (Paul Charles William),
1946– II. Brown, J.R. (Julian R.), *1957-*
539.7'54

Library of Congress Cataloguing in Publication data
Superstrings: A Theory of Everything?
Includes index.
1. Superstring theories. I. Davies, P.C.W.
II. Brown, J.R. (Julian Russell), 1957–
QC794.6.S85S869 1988 539.7'21 88-5020

ISBN 0 521 35462 5 hard covers
ISBN 0 521 35741 1 paperbackz

DQ

CONTENTS

PREFACE

In the last few years a remarkable new theory has captured the imagination of physicists. Known as string theory, or in its most developed form as superstring theory, it promises to provide a unified description of all forces, all the fundamental particles of matter, and space and time — in short, a Theory of Everything. The crux of the theory — that the physical world is made out of nothing but little strings — may sound absurd. But the theory is founded upon elegant mathematical ideas whose consequences are proving to be encouragingly consistent with the real world.

Such is the appeal of string theory that it has now become one of the liveliest research disciplines in theoretical physics, and has attracted the attention of many distinguished theorists. Some scientists have become greatly excited by the prospects of the theory, and have made spirited claims for its possible success. Nevertheless, the theory is not without its critics as you will discover by reading this book.

In 1987 we decided to review the state of superstring research by making a documentary on the subject for BBC Radio 3. The programme *Desperately Seeking Superstrings* was broadcast in early 1988. For this we sought the elucidation and opinions of some of the leading proponents and critics of string theory. As in our previous collaboration *The Ghost in the Atom* (Cambridge University Press, 1986) which started life as a radio documentary before metamorphosing into a book, we felt that it would be worthwhile publishing the interviews in a fuller and more permanent form. We have adhered to the original transcripts as much as possible, but inevitably some alterations were required to render the interviews more suitable for the printed page. We have, however, endeavoured to retain their conversational character.

The intention of this book is to give both physicists and interested non-physicists an insight into the essential ideas of string theory. We also hope that readers will find that the book offers a useful glimpse of how leading physicists talk and argue about a subject of contemporary importance. As the interviews were intended for a general audience, we asked our contributors to talk informally and to avoid using too much technical jargon. Each of the interviews is self-contained, and can be read in isolation. However, to provide a linking framework, we have written an extensive introduction, in which many of the background ideas that are needed for a fuller understanding of the theory are explained. There you will find brief accounts of quantum physics and the theory of relativity, as well as a survey of particle physics.

Although the subject of superstrings is still in a state of rapid development, the essence of the theory is now well established, and we hope that this book will provide a useful and entertaining snapshot of what could turn out to be one of the great scientific advances of our age.

We should like to thank Dr Ian Moss for assistance with the word processing and Miss Aileen Dryburgh for transcribing the audio tapes.

P.C.W. Davies
Julian R. Brown

1

Introduction

1.1 What is a Theory of Everything?

No science is more pretentious than physics, for the physicist lays claim to the whole universe as his subject matter. Whereas biologists are restricted to living organisms, chemists to atoms and molecules, psychologists to man and his fellow creatures, and so on, physicists, like theologians, are wont to deny that any system is in principle beyond the scope of their subject.

Physicists, of course, readily concede that in practice their understanding of most systems is woefully limited. Systems as basic as clouds and snowflakes are notoriously hard to model using the familiar laws of physics. As for biological systems, even the most primitive of organisms such as a virus or bacterium defeat the efforts of the physicist by virtue of their overwhelming complexity. Nevetheless, this practical impotence tends to be dismissed on the grounds that however mysterious a complex system may be, its behaviour must ultimately be dictated by the laws of physics, and *nothing* else.

The notion that nothing but the laws of physics are needed to account for the entire universe, in all its infinitely subtle detail, is encouraged by the philosophy of reductionism. Advocates of this school of thought, which includes many scientists, believe that, in principle, psychology can be reduced to biology, biology to chemistry and chemistry to physics. In other words, the 'arrows of explanation' always point downwards to the deepest layer of reality until, ultimately, everything can be explained in terms of the fundamental constituents of matter. The reductionist will therefore assert that if a consistent and complete theory of these constituents can be elucidated, it will *ipso facto* constitute a Theory of Everything.

We do not intend to argue here whether this reductionist line of reasoning can be sustained, but merely point out that it is in this spirit that some physicists have recently been talking about Theories of Everything, or TOEs. It is important to realize, however, that such a 'Theory of Everything' is not going to *explain* everything, just as in mathematics the axioms of geometry do not really 'explain' Pythagoras' theorem. It is true that Pythagoras' theorem can be deduced from the axioms, but the proof involves a fairly complicated chain of reasoning. The point is that even if we identify the fundamental elements of the physical world, we cannot expect an understanding of its many complex features to follow automatically. Thus the physicists' TOE will not solve many practical problems such as how to model clouds or snowflakes, let alone impinge on such grander mysteries as the origin of life or the nature of consciousness. Yet, according to the reductionist philosophy, an explanation for all of these phenomena should in principle be derivable from the TOE.

The first TOE (as far as we know) was constructed in the fifth century BC by the Greek philosophers Leucippus and Democritus. Their theory was called Atomism, and it asserted that the world consisted of nothing but atoms and the void. There were a number of different types of atoms, but all were supposed to be *elementary*, in the sense of being impenetrable and indestructible. This implied that atoms had no internal parts; they could not be said to be 'made up of' anything smaller. Atoms would have to be too small to observe directly and to be in a state of continual motion through the void. It was further argued that encounters between atoms might cause some to stick to others, producing the impression of continuous matter and that any change in the physical world could be attributed to the rearrangements of these atoms.

Following the rise of modern science with the work of Galileo and Newton in the seventeenth century, the atomic theory was supported by the elucidation of the laws of motion for material bodies. It now became possible to conceive that even the movement of the atoms obeys well-understood physical laws. This advance inspired Laplace to invent his famous demon calculator:

> An intelligence knowing, at any given instant of time, all forces acting in nature, as well as the momentary positions of all things of which the universe consists, would be able to comprehend the motions of the largest bodies of the world and those of the smallest atoms in one single formula, provided it were sufficiently powerful to subject all data to analysis; to it, nothing would be uncertain, both future and past would be present before its eyes.

This surely constituted an attempt at a Theory of Everything.

Some things, however, were conspicuously absent from this would-be TOE. No attempt was made to explain why the universe contains the atoms it does. The question of where they came from and why they have the masses and forms they have was left unanswered. The nature of the forces that act between atoms was also somewhat vague. Newton had provided a theory of gravity, but this was inadequate to account for all interatomic forces. Furthermore, the space through which atoms move, and the time by which their progress is gauged, lay outside the scope of the theory entirely. Space and time were regarded as simply *there,* not themselves part of physics. In these respects, then, the work of Galileo, Newton and Laplace could not be considered to constitute a very satisfactory TOE.

The situation remained largely unchanged until the latter half of the nineteenth century, when Newton's laws of mechanics and gravitation were supplemented with Maxwell's theory of electromagnetism. For a while it was possible to imagine that all natural forces might be a manifestation of either gravitation or electromagnetism in one guise or another. Although there was still no explanation for the existence of atoms, and space and time remained outside physics, many physicists assumed that their job from henceforth merely involved measuring the next decimal place of various physical quantities. In a speech to the British Association for the Advancement of Science in 1900, Lord Kelvin said: 'There is nothing new to be discovered in physics now. All that remains is more and more precise measurement.' There was thus a feeling abroad that a TOE was at hand.

With the benefit of hindsight we can see that an unsatisfactory feature of any purported TOE at that time was the need to postulate *two* fundamental forces: gravitation and electromagnetism. An attempt to remedy this shortcoming was made in about 1920 by the mathematician Theodor Kaluza, who discovered a possible link between the two forces (discussed in more detail later). Thus a serious candidate for a TOE might have come to prominence much earlier this century had physics not already been caught up in a conceptual maelstrom.

The discoveries of the electron and of radioactivity, the success of Planck's quantum hypothesis and the inception of Einstein's theory of relativity swept away the entire basis of Newtonian–Maxwellian physics. Newton's laws of motion and his commonsense assumptions about space and time were abandoned. Even Democritus' atomic hypothesis had to be replaced by a more subtle and complex view of the microworld in which atoms could no longer be regarded as indestructible particles with a well-defined position and motion. It became apparent that the foundations of classical physics had collapsed.

By about 1930 their place had been taken by new theoretical schemes: *quantum* mechanics, the general theory of relativity and a more elaborate model of the atom. Though many details remained obscure, physics seemed once more to be returning to a relatively simple set of principles. Although what had been called the atom turned out to be a composite body, it was still possible to conceive of all matter being composed entirely from a small number of *elementary* particles (electrons, protons, neutrons), subject to the laws of relativity and quantum mechanics. Indeed, in a mood of unrestrained optimism Eddington published as early as 1923 the beginnings of his so-called fundamental theory, an ambitious attempt at a TOE based on curious numerological relationships. Eddington continued to elaborate such ideas until his death in 1946. Einstein too spent much of his later years searching for a 'unified field theory' based on a purely geometrical description of nature.

But as we now know, all hopes for a TOE based on 1920s physics were premature. The need to postulate neutrinos, the discovery of the positron and the muon, and the clear manifestation of additional forces associated with the atomic nucleus eventually put paid to the idea that the fundamental rules of the universe operate at the level of simple interactions between electrons, neutrons and protons. Experimental *particle physics* flourished, unveiling a plethora of subatomic fragments and a bewildering tangle of forces. Physics proved to be far more complicated than had been suspected in the 20s.

It has taken a further half century for physicists to identify the deeper level of structure that underlies this subatomic richness, and to provide a more or less satisfactory theory of matter and forces. With this new understanding, some physicists have recently felt sufficiently emboldened to raise once more the prospect of a Theory of Everything. Superstring theory is the latest and by far the most promising attempt at a TOE. The regime in which this new confluence of ideas applies is ultra-microscopic. The world of the atom, or experimental high energy particle physics, is fully twenty powers of ten larger than the realm of the superstring.

What should we expect from a truly satisfactory TOE? First, it should explain why physicists observe the various elementary particles that they do, and correctly predict all of their key properties such as mass, electric charge, magnetic moment and so on. Second, it should faithfully describe all the interactions between the particles, which means that it should account for not only the four fundamental forces of nature, but also their relative strengths. Calculations with the theory also ought to yield precisely the observed values of the various inter-particle scattering amplitudes, decay rates, branching ratios, etc. In short, the theory should account for all the measured parameters of particle physics. In addition to this, it should provide an explanation for the geometry and topology of spacetime, such as the number of perceived dimensions, and offer a convincing account of how the universe came into existence.

But this is not all. A TOE should also *unify* physics.

1.2 Unity at the heart of nature

Anybody can have a stab at constructing a TOE. Just go to the textbooks, write down all the basic laws, ennumerate all the known subatomic particles and their forces, and display the whole lot. There it is, everything you always needed to know about the universe!

So what is wrong with that approach? The problem is partly aesthetic: such a list looks messy. A good TOE should consist of much more than a mere catalogue of underlying laws and objects; it should have explanatory power, and it should establish *linkages* between the various facets of nature. Admittedly, the search for such a TOE is to a certain extent an act of faith, motivated by the deep belief that nature ought to be simple.

Generally speaking, the fewer the number of independent assumptions in a scientific theory, the more powerful and compelling it is. Theories often retain free parameters which have to be fixed empirically. With further advances, a deeper theory might then provide the values of those parameters. To take a specific example, the Stefan–Boltzmann radiation law of the 1880s related the energy density of black body radiation to the fourth power of the temperature. The constant of proportionality was fixed by experiment. Following Planck's successful quantum hypothesis in 1900, this quantity was shown not to be an independent fundamental constant of nature, but derivable in terms of other constants of physics, viz. the speed of light, Planck's constant and Boltzmann's constant.

Scientific experience has demonstrated many times that the deeper one probes, the more connections become apparent and the less one needs to fix apparently arbitrary aspects of theory by appeal to direct experiment. Thus the modern concept of the atom improves on the classical picture because it shows how different types of atoms arise from the same few constituents of matter. There is no need to postulate ninety or so different fundamental objects, each corresponding to a chemical element. Today, the physical and chemical properties of different atoms can be related systematically in terms of their constituent parts.

The ultimate TOE would, ideally, need no recourse to experiment at all! Everything would be defined in terms of everything else. Only a single undetermined parameter would remain, to define the scale of units with which the elements of the theory are quantified. This alone must be fixed empirically. (In this ultimate case, experiment merely serves to define a measurement convention. It does not determine any parameter in the theory.) Such a theory would be based upon a single principle, a principle from which all of nature flows. This principle would presumably be a succinct mathematical expression that alone embodied all of fundamental physics. In the words of Leon Lederman, director of Fermilab – the giant particle acclerator facility near Chicago – it would be a formula that you could 'wear on your T shirt'.

Attempts to unify physics follow two quite different strategies. The first of these might be called the 'top-down' approach. One starts with some broad, overarching principle, perhaps justified on the grounds of elegance or simplicity and probably expressed in succinct mathematical form. One then works towards a description of the world, arriving only in the last analysis at specific predictions.

Much of Einstein's work exemplified the power of the top-down route. His development of the general theory of relativity was founded on the principle of the equivalence of gravitational and inertial forces, and the principle that physics must be independent of the coordinate system in which events are described. From such basic and elementary ideas Einstein was led almost uniquely to his gravitational field equations. The equations themselves are renowned for their elegance, simplicity and compactness. However, the *solution* of these equations is usually far from simple. Computations of, say, the motions of the planets or the emission of gravitational radiation from binary stars are very complicated. So even after more than sixty years, the full consequences of the theory have yet to be revealed.

In contrast is the 'bottom-up' approach to science which is probably more commonplace. Here one starts out with phenomenology. Laboratory experiments provide a mass of raw data, which is then organized systematically, and certain regularities deduced. These regularities are refined to specific postulates, leading to laws of greater generality. Predictions are made about phenomena in new areas, and experiments used to test those predictions. Piece by piece, the scientist fashions linkages here and there, and with any luck he eventually assembles a theory that is greater than the sum of its parts. Particle physics provides many examples of the success of this approach. The quark theory, for instance, was arrived at by successive stages of unification of apparently different particles into families, or multiplets, suggested by various physical similarities deduced from experiment. The essence of the bottom-up route is that it never strays too far from the path indicated by experiment, and if the experimental facts lead away from philosophical elegance, so much the worse for philosophy.

The history of physics is the history of successive stages of unification. Newton, for example, demonstrated that the motion of the heavenly bodies conforms to the same dynamical and gravitational laws as bodies near the surface of the Earth. Maxwell unified the laws of electricity and magnetism, and in addition established a link between electromagnetic field theory and optics by showing that light consists of electromagnetic waves. Einstein found a connection between space and time, and energy and mass, and then went on to link spacetime to gravitation.

In recent years attempts to unify nature further have focused upon high energy particle physics. The reason for this is that the deeper the level of structure we want to examine the higher the energy we have to use. In this regime, two theories are paramount. The first is the theory of relativity, the second is quantum theory. A starting assumption of all recent attempts to unify physics is that these two theories must be explicitly incorporated. Before we examine particle physics, then, let us briefly review these two theories.

1.3 The theory of relativity

In the late nineteenth century it became apparent that Newton's laws of mechanics and Maxwell's theory of electrodynamics gave conflicting descriptions of relative motion. According to Galileo and Newton, uniform motion in a straight line is purely relative to some frame of reference; the motion does not produce any absolute physical effects. Thus in a smoothly flying aircraft the passenger has no real sensation of speed, and can verify his motion only by reference to some external system, e.g. by looking out of the window and seeing the ground go by. On the other hand Maxwell's equations predict that electromagnetic waves such as light travel at a velocity which is a fixed constant for free space. The speed of light is therefore a universal constant of nature. The theory says nothing about the frame of reference against which this speed should be measured. An idea developed that space must be filled with an invisible medium called the *aether* through which light propagates. The aether would then define a universal frame of rest against which all motion could, in principle, be gauged. This it seemed would reconcile Maxwell's theory with the principle of relativity enunciated by Galileo and Newton. However, experiments to determine the speed of the Earth through the aether gave the result zero. This was utterly paradoxical, for it seemed to imply that the Earth was at rest and the heavens in motion about it!

In 1905 Einstein published his so-called special theory of relativity which addressed this paradox directly. Einstein upheld the principle of relativity – that uniform motion in a straight line is purely relative – but rejected the aether. Instead he introduced a new postulate into physics. This states that the speed of light is the same in all reference frames. It implies that, irrespective of the state of motion of the light source or observer, light will pass the observer at exactly the same speed. Two observers in relative motion who observe the *same* light pulse will nevertheless measure the *same* speed for that pulse.

Einstein's new principle of the constancy of the speed of light requires one to discard commonsense notions of space and time. In particular, Newton's treatment of space, time and motion had to be replaced with a new 'relativistic' theory. Central to the relativistic concept of space and time is that spatial distances and intervals of time will be perceived differently by different observers according to their state of motion. In this way an interval of, say, one hour on Earth might be measured as only half an hour in an incredibly fast spacecraft. Similar remarks apply to lengths. These relativistic effects, however, are very small unless the speeds involved are close to that of light. In high energy particle physics the particles often travel very close to the speed of light and experience very pronounced time dilation. The effect shows up directly in the way in which unstable particles take much longer to decay (as measured at the laboratory bench) when whirling around a particle accelerator than they do at rest.

These distortions of space and time have a profound effect on the laws of mechanics. For example, a moving body is, roughly speaking, heavier than when at rest; its mass rises as it is accelerated. The concept of mass is therefore somewhat ambiguous. Physicists usually define the mass of a body to be its mass as measured when at rest (relative to the observer making the measurement). The effective mass (or relativistic mass) as represented by, say, the body's inertia, depends on its speed, and rises without limit as the speed of light is approached. An examination of the relationship between mass, speed and energy in the theory of relativity reveals that mass and energy are equivalent. The statement means that energy has mass, and that matter is a form of energy. It also implies that, under certain circumstances, matter might be destroyed and replaced by some other form of energy. Conversely, energy might be used to create matter. These ideas are embodied in Einstein's famous equation $E = mc^2$, where E is energy, m is mass and c is the speed of light.

As a consequence of these ideas, it is impossible for a particle, or any other object, to 'break the light barrier'. That is, no particle can accelerate from less than the speed of light to greater than the speed of light. One way to understand this is in terms of mass, for the mass of an object would become infinite at the speed of light and thus require an infinite amount of energy to accelerate it to that state, which is impossible.

The above restriction does not, of course, imply that nothing can travel at the speed of light, for clearly light itself does. For a particle to travel at the speed of light its rest mass must be zero (though not its actual mass, for such a particle can never be at rest). Moreover, the theory of relativity does not specifically exclude the possibility of particles that travel faster than light. Such hypothetical particles have gained the name *tachyons*;

these are forbidden by the theory of relativity from crossing the light barrier the other way, i.e. they can never travel slower than light. According to the equations of relativity a tachyon's rest mass would be an imaginary number (i.e. the square root of a negative number). This would be an embarrassment if the rest mass of tachyons were a measurable quantity, but as they are forbidden from travelling slower than light they could never be brought to rest anyway.

Although tachyons are allowed by the theory, most physicists regard them with distaste. For a start, their faster-than-light motion implies that under some circumstances they could travel backwards in time. If tachyons can interact with ordinary matter, it seems that they could convey messages to the past, thus leading to the possibility of all sorts of awkward causal paradoxes. Some attempts have been made to circumvent this by redefining the direction of time along the paths of such tachyons (i.e. by regarding a tachyon travelling backwards in time from place A to place B as a tachyon travelling forwards in time from B to A) but it is not clear whether this can be done consistently. The idea of tachyons runs into even greater problems when quantum theory is taken into account.

The transformation of space and time intervals predicted by the theory of relativity implies that space and time are part of physics, rather than an arena in which physics takes place. Indeed, the way in which the transformations work reveals that space and time are inextricably linked, and should be regarded as forming a four-dimensional continuum called *spacetime*. For this reason physicists think of the world as four-dimensional rather than three-dimensional.

It soon became clear to Einstein that the special theory of relativity meant abandoning not only the Newtonian concepts of space and time and the laws of mechanics, but also Newton's theory of gravity. According to Newton, gravitational forces act instantaneously across space. But this would violate the theory of relativity because it implies that gravitational effects travel faster than light.

Einstein set out to construct a new theory of gravitation based upon a generalization of his theory of relativity. This took many years; he completed the task in 1915. In the original 'special' theory, Einstein was concerned with uniform motion.

If a body accelerates, its motion is no longer merely relative to some reference frame. An aircraft passenger, for example, will certainly feel motion if the plane suddenly banks or descends. In order to encompass this more general type of motion, Einstein took into account that an acceleration produces forces which are indistinguishable from gravity. Thus a centrifuge is sometimes described as producing 'artificial gravity'. One talks of 'g forces' acting during the rapid acceleration of a space rocket.

This equivalence of acceleration and gravitation was known to Galileo and Newton, but Einstein elevated it to a fundamental principle of nature. It is usually discussed in connection with falling bodies. If a body is allowed to fall freely, the 'g force' due to its downward acceleration exactly balances its weight. An observer in free fall therefore experiences weightlessness. These days this condition is familiar to astronauts in orbit, but in Einstein's day he chose to imagine an observer in a freely falling elevator.

Because the contents of a freely falling elevator are effectively weightless, neighbouring bodies within the elevator do not change their relative positions. Viewed from the reference frame of an observer standing on the ground, these bodies all fall with equal acceleration, and it is this fact which motivates the statement that objects which are dropped together strike the ground together, irrespective of their weight or constitution. (It is said that Galileo demonstrated this fact by dropping balls from the leaning tower of Pisa.)

Strictly speaking, the foregoing remarks are only true to the extent that we neglect three things. The first is air resistence, which is an irrelevant complication. (The experiment works better on the moon which has no atmosphere.) The second is the minute force of gravitational attraction between the bodies within the elevator. This can be made arbitrarily small by assuming that the bodies are very light so that their gravitational power is negligible. Such bodies, which merely trace out the gravitational field in which they are situated without contributing significantly to it themselves are referred to as 'test' bodies.

The third factor which was ignored above is the curvature of the Earth. Although this is very small in practice, its effect is absolutely crucial to an understanding of gravity. To see why, look at Figure 1, which depicts schematically an elevator populated by two test bodies, falling freely towards a spherical Earth. If the curvature of the Earth is ignored, the bodies are seen to fall on exactly parallel paths, so that they do not change station relative to each other. In fact, each test body is pulled directly towards the Earth's centre, so that they fall on very gradually converging trajectories. An observer incarcerated within the box, unable to view the outside world, would be able to deduce the Earth's curvature from observations of the rate of convergence of the test bodies.

Figure 2(*a*) depicts the situation in which four test particles are arranged in a diamond shape. As the elevator falls the middle pair converge as before. The body at the base of the

Figure 1. Two nearby freely falling particles slowly approach one another as they drop towards the geometrical centre of the Earth on gradually converging trajectories.

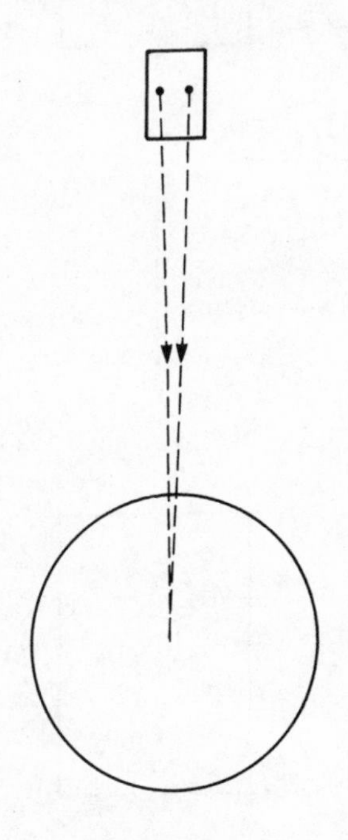

diamond, being slightly closer to the Earth than the others, experiences a slightly stronger pull, because the strength of the Earth's gravity diminishes with distance according to the inverse square law. Thus the lowest body falls slightly faster than its neighbours. For similar reasons the top body falls

Figure 2(a). Four particles arranged in a square fall freely. (b) Differential gravitational forces gradually distort the square to a diamond shape. The bottom particle is nearer the Earth, so feels a slightly stronger gravity and falls more rapidly. The top particle lags behind all the others as it is furthest from the Earth. The outside pair fall slowly together as explained in Figure 1

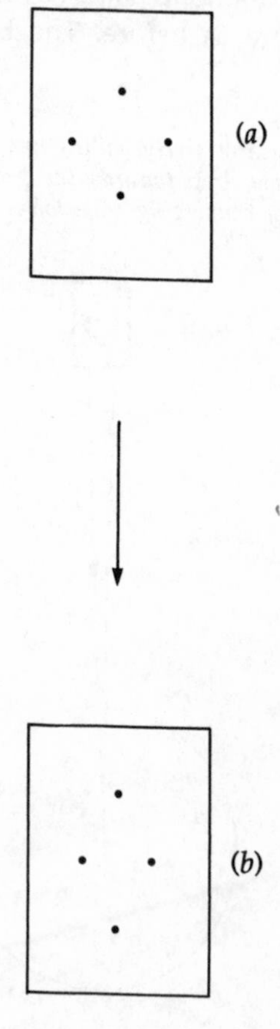

slightly slower. The upshot is that while the diamond is compressed horizontally, it is stretched vertically, leading to the more elongated shape shown (exaggeratedly) in Figure 2(*b*).

The elevator experiment shows that by going into a state of free fall one has found a reference frame in which the gross effect of gravity is abolished. Nevertheless, so long as the gravitational field isn't uniform, gravity is still manifested through the slight displacements of test bodies. These subtler gravitational influences are referred to as 'tidal' forces because they are responsible for producing the ocean tides in the field of the moon's gravity.

One might say that gross gravitational forces are merely relative to one's reference frame, but tidal forces are absolute and represent the real gravitational field. It was therefore the tidal field that Einstein sought to model in his general theory of relativity. The key feature is that the distortion of the geometrical shapes such as the diamond is independent of the constitution of the test bodies or their masses (so long as the mass doesn't get so large that they no longer qualify as test bodies). This suggests that the distortions concerned should be regarded as an underlying property of the space through which the bodies fall, rather than as a result of forces which act on the bodies. In other words, one can envisage the bodies to be falling freely through a distorted or warped space rather than being acted upon by forces. Thus Einstein was led to the idea that gravity might be nothing more than geometry – a distortion in the geometry of space.

Let us examine this idea in more detail. First, an important point: the theory of relativity links space and time, and it is actually a distortion of *spacetime* rather than space alone which is relevant here. (A distorted spacetime may or may not imply a distorted space.) At school we learn the geometry of Euclid, which is adapted to flat sheets and, in three dimensions, flat space. On a curved surface the rules of geometry differ, as may be illustrated with the help of Figure 3. On the surface of a sphere it is impossible to draw parallel lines, for example. The analogues of straight lines on a spherical surface are great circles, such as lines of longitude. Two such lines are shown. They start out parallel at the 'equator', but intersect at the 'north pole'. This distortion of paths or lines on a curved surface is similar to the distortion of the paths of falling test bodies in a non-uniform gravitational field. The main difference in the latter case is that the distorted geometry is not two-dimensional (or even three-dimensional): it involves three dimensions of space and one of time. Although it is very hard to actually visualize curvature in four dimensions, it is a straightforward matter to describe it mathematically.

Einstein's general theory of relativity treats the gravitational field as a field of geometrical distortion, a curvature or warping of spacetime. In this theory freely falling bodies are not treated as subject to gravitational forces, but are instead regarded as following the straightest possible path (known as a geodesic) in an underlying curved spacetime. In Newton's theory of gravitation, the Earth's orbit curves around the Sun because the Sun's gravity forces it to depart from its natural straight line motion. In Einstein's theory there is no gravitational force as such (though we shall continue to refer loosely to the 'force of gravity').The Sun produces a warping of spacetime in its vicinity and the Earth travels freely along a geodesic in this curved spacetime. Gravity is treated as a geometrical effect

Figure 3. Curved geodesics. Because the geometry of the Earth is not flat, two 'straightest' paths (geodesics) that are parallel at the equator, converge and intersect at the north pole. This is analogous to the tidal forces that cause two particles to drift together in a falling elevator.

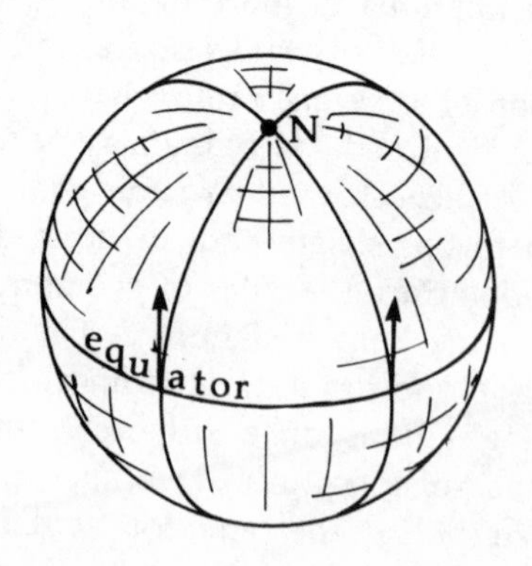

precisely because it is universal; it affects all test objects in the same way. Thus, even light will follow a curved path in a gravitational field. Figure 4 shows the effect of the Sun's gravity on a passing starbeam, which is measurably bent. On a larger scale, the distribution of galaxies throughout the universe will depend on the geometry of space.

The fact that there might be a systematic curvature of space on a cosmological scale raises the interesting question of the *topology* of the universe. So long as space is considered to be flat, it must be either infinite in extent, or else possess some sort of boundary. But if space is curved there are other possibilities. Think of the situation with a two-dimensional sheet. A curved sheet could be closed into a sphere, for example, or a torus (see Figure 5). (Remember, although a sphere is a three-dimensional object its surface is only two-dimensional.) It is possible to envisage a three-dimensional version of the closed spherical surface, called a hypersphere. If the universe had the topology of a hypersphere it would possess a finite volume, but there would be no boundary or edge to space. It is not known what topology space actually possesses, but the issue is crucial to the superstring theory, as we shall see.

Figure 4. Light is bent by gravity. The Sun's gravity bends the light beam so that the distant star A appears from Earth to be shifted to position B. This shift can be observed and measured during an eclipse, when the moon conveniently blocks out the Sun's glare, allowing the stars to be seen in daytime.

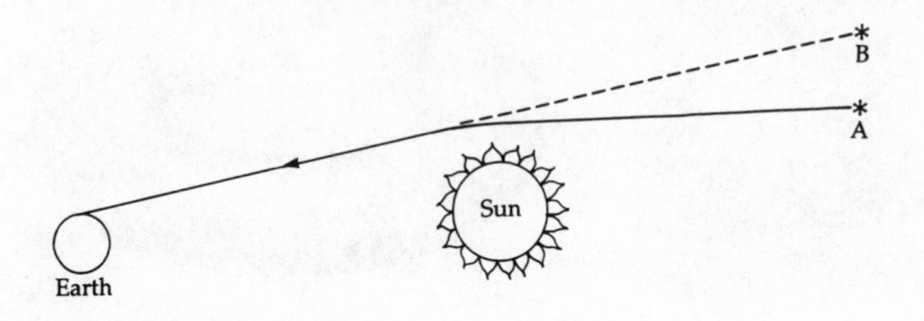

1.4 The quantum theory

The theory of relativity requires us to relinquish some cherished notions about space, time and motion. It replaces the intuitive physics of Newton with an altogether more abstract picture, involving concepts such as warped spacetime that may be hard or even impossible to visualize. The quantum theory demands an equally radical reappraisal of commonsense ideas about the nature of matter.

Quantum theory began in 1900 with the suggestion by Max Planck that electromagnetic radiation comes only in discrete packets, or quanta, which we now call photons. A photon can be regarded in some sense as a particle of light. This idea is hard to reconcile with the traditional assumption that light and other forms of electromagnetic radiation consist of waves. The apparent contradiction is resolved by the concept of *wave–particle duality*, according to which light can manifest itself in both wave-like and particle-like forms depending on the way in which it is observed, i.e. the particular experiment being performed. Light cannot, however, behave as *both* a wave and a particle at the same time.

Figure 5. Topology is the study of the way that lines, surfaces, etc. connect to themselves. The topology of the sphere (left) differs from that of the torus (right) because the torus contains a hole.

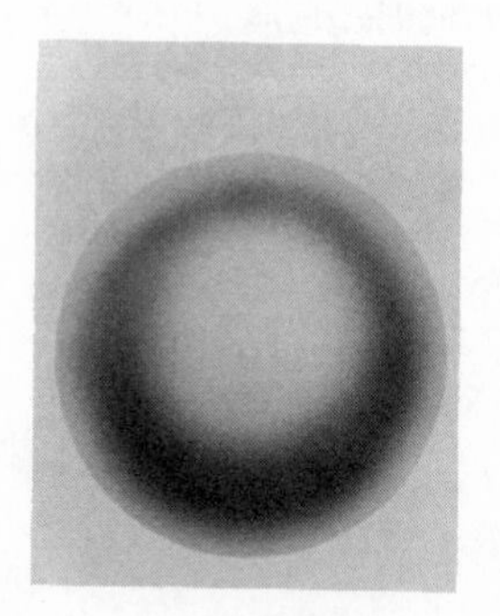

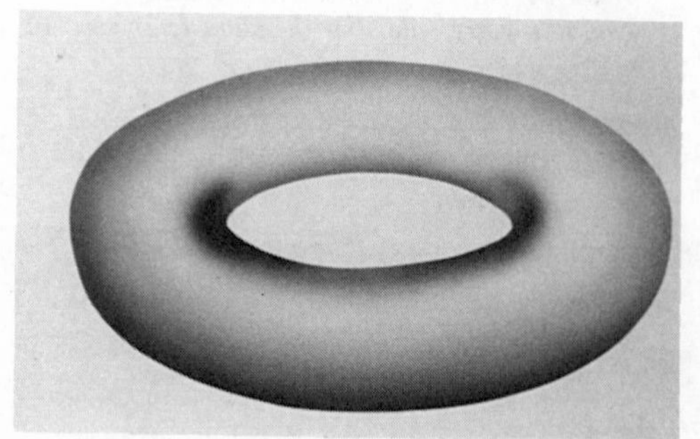

Niels Bohr described this state of affairs by calling the wave and particle forms *complementary* aspects of a single reality, a reality that is beyond our ability to visualize. It is conversely the case that electrons, protons and other subatomic objects, which are normally regarded as particles, can appear as waves under some circumstances. Thus the photon has the same general status as these entities, and can be placed alongside them as another species of particle.

The wave–particle duality that is central to the quantum theory implies that some of the familiar qualities one would wish to attribute to a subatomic object may not always be well defined. A uniform quantum wave, for example, has a definite momentum, but being extended in space it has no well-defined position. On the other hand if, say, an electron or photon is measured in such a way as to reveal its position (by using a photosensitive screen for example), its momentum becomes ill-defined. Thus one can perform one sort of measurement to determine the momentum, and another sort of measurement to determine the position, but these two measurements are mutually exclusive: one can never simultaneously determine both the position and the momentum of a quantum object. As a result, there is an irreducible fuzziness or vagueness in the activity of a quantum system, a fuzziness that is represented by the famous uncertainty principle of Werner Heisenberg.

One way of stating the uncertainty principle is to say that all measurable quantities are subject to unpredictable fluctuations which smear out their values. Quantities are associated in incompatible pairs, such as position and momentum. If the degree of uncertainty, or smearing, of position is denoted by Δx and that of momentum by Δp, then the uncertainty principle demands that the product $\Delta x \Delta p$ can never be less than a universal constant called Planck's constant and denoted by h. Thus h quantifies the degree of fuzziness in nature. The numerical value of h is very small (6.63×10^{-34} joule seconds), reflecting the fact that quantum fuzziness is only conspicuous in the atomic and subatomic realm. In principle, however, it applies to all systems.

Another important pair of quantities to which the uncertainty principle applies is energy, E, and time, t. In this case $\Delta E \Delta t$ cannot be less than h. Taken together, these two versions of the uncertainty principle imply that the uncertainty in position can only be made very small by having a large uncertainty in momentum, while the uncertainty in time can only be reduced at the expense of a large uncertainty in energy. In many cases this is equivalent to the statement that to discern structure on very short length and time scales requires the availability of very large momenta and energy. Quantum theory thus attaches a natural energy and momentum scale to each interval of time and space. It follows that the smaller the region a physicist chooses to explore, the larger the energy (and consequently momentum) needed to do so. For this reason, large particle accelerators are needed to investigate small structures. Moreover, any theory of the ultimate structure of matter, which necessarily refers to the smallest length scales, involves the largest energies. Special interest therefore attaches to the *high energy* characteristics of such theories.

Because of the irreducible uncertainty inherent in quantum systems, the Newtonian laws of mechanics fail (even if relativistic effects are taken into account) for objects such as electrons, and have to be replaced by an entirely new *quantum mechanics*. This was developed in the 1920s by Heisenberg, Erwin Schrödinger and others. Similarly, the dynamical equations for fields, such as Maxwell's equations for the electromagnetic field, have to be replaced by a new quantum theory of fields. Work on this began in the 1930s.

Subatomic particles often move close to the speed of light, and an important requirement is that their quantum description is consistent with the special theory of relativity. This development of quantum mechanics was made in 1929 by Paul Dirac, whose relativistic treatment led to the successful prediction of antimatter (see below). Furthermore, when quantum theory is applied to fields (such as the electromagnetic field) a mathematically consistent theory can be obtained if it is cast in a relativistic form. Even so, the theory ran into severe mathematical difficulties, and it took until after the Second World War for a successful relativistic quantum field theory, called quantum electrodynamics (QED for short) to be constructed. All current attempts to produce a fundamental theory of physical phenomena take it for granted that they must be consistent with both the special theory of relativity and with quantum mechanics.

1.5 The world of subatomic particles

An inventory of all the known subatomic particles runs into hundreds. The electron, proton and neutron are merely three among a zoo of objects. The others are found in cosmic rays or are made in particle accelerators by colliding other particles at very high energy. All but a handful of subatomic particles are highly unstable, and decay into other particles in a minute fraction of a second.

All members of a given particle species are identical; there is no way to distinguish, say, one electron from another. Moreover, every species of particle has an associated antiparticle in which all the distinguishing qualities except mass are reversed. Thus the electron possesses a fixed quantity of negative electric charge and the antielectron (better known as the positron) has the same quantity of positive charge.

Particles are characterized by a range of physical qualities. Mass and electric charge are two of the more important. For some deep reason, as yet uncertain, the electric charge on all known particles comes in simple multiples of the fundamental unit of charge carried by the electron. On the other hand the masses of the different particles bear no simple numerical relation to one another.

Another very important characteristic of subatomic particles is their intrinsic spin. Many particles possess a type of internal rotation that for some purposes can be envisaged as akin to a body spinning on its axis. Actually, spin is a quantum mechanical property, and it has no direct analogue in classical mechanics. To give an example of the way in which intrinsic spin differs from the ordinary spin of a body such as the Earth, consider the question of the magnitude of the angular momentum. The angular momentum of a macroscopic body can assume any value from a continuous range. In the case of a subatomic particle, however, the angular momentum is 'quantized', which is to say that spin always comes in fixed discrete units that are integer multiples of $\frac{1}{2}\hbar$, where $\hbar$ is

Planck's constant h divided by 2π. For ease of expression, one talks of a particle with spin $\frac{1}{2}\hbar$ as having 'spin $\frac{1}{2}$'. Thus the electron has spin $\frac{1}{2}$, the photon has spin 1, the so-called Ω^- particle has spin $\frac{3}{2}$, and so on.

There are also oddities concerning the geometrical properties of intrinsic spin. If an ordinary spinning body is rotated in space through 360° it returns to its original configuration. A particle with spin $\frac{1}{2}$, however, will not do this. If such a particle is rotated through 360° it assumes a quantum state with measurably different physical properties. To return the particle to its initial state it is necessary to rotate it through 720°. In other words, a spin $\frac{1}{2}$ particle requires a double rotation relative to 'everyday' objects before it 'comes back to its starting state'. It is as though a spin $\frac{1}{2}$ particle somehow sees a bigger world than we do. Our view of space is double-valued relative to that of the particle. What to us are two identical copies of the universe, one for each 360° rotation, are to the particle distinct. Clearly the geometry of space is fundamentally and subtly different for a spin $\frac{1}{2}$ particle.

It turns out that the precise value of a particle's spin also plays a crucial part in determining its physical properties. Particles endowed with an even number of spin units ($\frac{1}{2}\hbar$) behave quite differently from those those with an odd number. The former are referred to as *bosons*, and the latter as *fermions*. Fermions are subject to the Pauli exclusion principle, which states that no two identical particles can occupy the same quantum state. No such restriction applies to bosons.

The building blocks of matter can be divided into two other distinct classes. The first are the so-called *leptons*, meaning 'light ones'. The most familiar lepton is the electron. There is another particle called the muon that differs from the electron in being some 206 times heavier. Muons are also unstable, and decay into electrons in about 2 microseconds. There is another, still heavier version of the electron called the tauon, which was discovered in the 1970s. It too is very unstable.

In addition to the three charged leptons, there are (probably) three electrically neutral particles known as neutrinos. Each species of neutrino is associated in its behaviour with one of the charged leptons. Thus there are electron-neutrinos, muon-neutrinos and, it is believed, tauon-neutrinos (tauon-neutrinos have yet to be detected). For a long time it was supposed that neutrinos have zero rest mass and travel at the speed of light. The mass of the electron-neutrino is certainly very small but there is no compelling theoretical reason why neutrinos have to be massless. At this stage nobody knows precisely what their masses are.

The six leptons are all fermions, with spin $\frac{1}{2}$. They are distinctive in being relatively weakly interacting: they do not take part in nuclear interactions. In contrast, nuclear particles interact fiercely. There are also many dozens of nuclear-associated particles in addition to the familiar proton and neutron. Collectively, the nuclear particles, and those particles made as a result of nuclear interactions, are known as *hadrons*.

Hadrons are generally much heavier than leptons; the proton, for example, is 1836 times as massive as the electron. The heavier hadrons tend to be fermions. Fermionic hadrons are given the collective name of *baryons*, meaning 'heavy ones'. There are also hadrons which are bosons. These are collectively called *mesons*, after their intermediate mass. The proton and neutron are baryons, and both have spin $\frac{1}{2}$. The lightest meson is the pion, which has spin 0. Table 1 shows some of the more common hadrons; most of them are known only by their Greek names. Of all the hadrons, only the proton is stable (and perhaps not even that — see Section 1.10). The rest decay either into lighter hadrons, or leptons.

The proliferation of hadrons suggests that they are not elementary particles, but composite bodies with internal parts. This is in contrast to the leptons, which are usually assumed to be fundamental. In the early 1960s Gell–Mann and Zweig proposed that hadrons are made out of smaller components called *quarks*. The quark theory is now well established.

Like the leptons, quarks come in (probably) six different varieties, whimsically known as *flavours*. These carry the arbitrary names up, down, strange, charm, top and bottom (or truth and beauty). Also like the leptons, all quarks have spin $\frac{1}{2}$, so they are fermions.

Table 1. *Some common hadrons*

Name	Symbol	Mass	Charge	Spin	Lifetime
Pion	$\pi^+ \pi^-$	139.57	+1 −1	0	2.6×10^{-8}
	π^0	134.96	0	0	0.8×10^{-16}
Kaon	$K^+ K^-$	493.67	+1 −1	0	1.2×10^{-8}
	$K^0 \bar{K}^0$	497.67	0	0	0.9×10^{-10}
Eta	η	548.8	0	0	2.5×10^{-19}
Proton	$p \bar{p}$	938.28	+1 −1	$\frac{1}{2}$	$>10^{39}$
Neutron	$n \bar{n}$	939.57	0	$\frac{1}{2}$	898
Lambda	$\Lambda \bar{\Lambda}$	1115.60	0	$\frac{1}{2}$	2.6×10^{-10}
	$\Sigma^+ \bar{\Sigma}^+$	1189.36	+1 −1	$\frac{1}{2}$	0.8×10^{-10}
Sigma	$\Sigma^0 \bar{\Sigma}^0$	1192.46	0	$\frac{1}{2}$	5.8×10^{-20}
	$\Sigma^- \bar{\Sigma}^-$	1197.34	-1 +1	$\frac{1}{2}$	1.5×10^{-20}
Xi	$\Xi^0 \bar{\Xi}^0$	1314.9	0	$\frac{1}{2}$	2.9×10^{-10}
	$\Xi^- \bar{\Xi}^-$	1321.3	-1 +1	$\frac{1}{2}$	1.6×10^{-10}
Omega	$\Omega^- \bar{\Omega}^-$	1672.5	-1 +1	$\frac{1}{2}$	0.8×10^{-10}

Masses are in MeV (millions of electron volts), charges in units of the proton charge and lifetimes in seconds. Where the antiparticle differs from the particle a separate symbol is given.

Quarks combine together to make hadrons in two different ways. One way involves the union of three quarks. According to the rules of quantum mechanics, the spins of the quarks have to be either parallel or antiparallel to each other, so three spin $\frac{1}{2}$ quarks together make up a body with total spin of either $\frac{1}{2}$ or $\frac{3}{2}$. These are the baryons. Depending on the various quark flavours, so the different combinations make up all the known baryons. For example, the proton has two up and one down quark, the neutron has two downs and one up, and the Ω^- has three strange quarks.

In the other way of combining together, a quark is joined with an antiquark. The rules of quantum mechanics then demand that the total spin be either 0 or 1, i.e. the result is always a boson. These correspond to the mesons. Because mesons only contain two quarks, whereas baryons contain three quarks, the mesons are generally lighter. However, the mass of the charmed quark, for example, is much greater than that of the up and down quarks, so that a meson made from a charmed quark–antiquark pair is actually considerably heavier than the three-quark proton.

As quarks combine in threes to make baryons, they must possess electric charges of either $\frac{1}{3}$ or $\frac{2}{3}$ of the fundamental unit (i.e. the charge on the proton). Such a fractional charge would make a solitary quark stick out like a sore particle if it could be observed experimentally. However, it is almost certain that quarks cannot be isolated. There is strong evidence that they remain locked in permanent union within hadrons. All attempts to smash hadrons by collision into their constituent quarks have failed, and from what is known about the inter-quark force (see below) it appears likely that quarks are indeed totally confined.

In spite of the fact that physicists have been unable to study quarks in isolation, convincing indirect evidence for their existence within hadrons comes from experiments in which electrons are shot through nuclear particles. The scattering pattern of these electrons reveals the presence of three massive compact bodies within each nuclear particle. Other evidence for quarks comes from the decay of hadrons, from the production of hadron 'jets' in high energy collision experiments, and elsewhere.

Most physicists are content to suppose that quarks and leptons represent the bottom level of structure, i.e. they are *the* fundamental particles from which all matter is built. It is of course conceivable that these particles are themselves made of yet smaller entities. Indeed, some physicists feel that the total number of quarks and leptons is embarassingly large. (In addition to the six quark flavours, each flavour comes in three different 'colours', making eighteen distinct quark types in all.) But accepting that quarks and leptons are, in fact, fundamental raises the question of what form they take.

To be fundamental, an entity cannot by 'pulled apart' or turned into something else by internal rearrangement. For this reason quarks and leptons have long been thought to be point-like, with no internal structure at all. However, as we shall see, there are severe theoretical problems associated with point-like particles, and it seems likely that these 'fundamental' particles actually have some sort of structure after all.

1.6 The four forces

Although in daily life nature seems to display a wide variety of forces, they can, in fact, all be reduced to just four. The most familiar is gravity, and this was the first force to receive a systematic mathematical theory — by Newton. Gravity is the only *universal* force; i.e. it acts between all particles without exception. The source of gravitation is the mass of the particle concerned, so gravity is a cumulative force, growing in strength as more and more matter is assembled. Except under some exotic circumstances, gravity is always attractive.

Gravity is termed a 'long-range' force, because it can act over macroscopic — indeed cosmological — distances. This is because the force falls off with distance relatively slowly — to be precise it falls off as the inverse square of the distance. The absolute strength of gravity is exceedingly small. The gravitational force between an electron and a proton, for example, is some forty powers of ten weaker than the electrostatic force. For this reason, gravity does not seem to play a very direct role in subatomic particle physics. Nevertheless, it is one of nature's fundamental forces, and a place must be found for it in any unified theory.

An important concept in describing all forces is that of a field. Newton conceived of gravitation as 'action at a distance', in other words one particle acts directly upon another across space. The more modern view is that each particle is the source of a field of force — in this case the gravitational field — which surrounds it. Another particle, on finding itself immersed in this field, experiences a force in proportion to the strength of the field at that point. The decline of gravity with distance is then attributed to the weakening of the field away from its source.

In 1915 Einstein replaced Newton's theory of gravitation by his general theory of relativity. As we discussed in Section 1.3, in the new theory, the gravitational field is interpreted as a distortion or curvature of spacetime, i.e. it is ascribed a purely geometrical nature. This geometrical interpretation sets gravity apart from the other forces.

After Newton's work on gravity, the next force that received a theoretical basis was electromagnetism. Electric and magnetic forces are easily demonstrated in the laboratory and have been known since antiquity. But it was not until the nineteenth century that Michael Faraday and others revealed an intimate link between electricity and magnetism. James Clerk Maxwell then succeeded in formulating a set of equations that unified the two into a single electromagnetic theory. This was the first definitive step towards a unified theory of the forces of nature.

The source of the electromagnetic field is electric charge. Not all particles are charged, so electromagnetism, unlike gravity, is not a universal force. It does resemble gravity, however, in being long-ranged — electric and magnetic forces obey an inverse square law analogous to the gravitational one. The strength of the electromagnetic force is, as already remarked, enormous compared to gravity, but because electric charge can be both positive and negative, there is a tendency within massive objects for cancellation to occur; the force is not cumulative, but self-neutralizing. For this reason gravity rather than the intrinsically much stronger force of electromagnetism dominates the universe on the large scale.

The two remaining forces are not directly observable in daily life because their ranges are of subnuclear dimensions. The first of these forces, called the strong nuclear force, is responsible for binding the protons and neutrons together in the nucleus. Beyond a distance of about 10^{-15} metres the strong force rapidly dwindles away. Its short-ranged character is in sharp contrast to the long range of gravity and the electromagnetic force. Not only protons and neutrons, but all hadrons, feel the strong force. Leptons, however, do not.

The form of the force between hadrons is very complex. This is because hadrons are not elementary particles but groups of quarks, and it is the inter-quark force which is the fundamental interaction. In essence, this force resembles electromagnetism athough it is much stronger. The complication arises in that unlike electromagnetism, which is a two-body force, the strong force is responsible for binding three quarks together in baryons. This calls for a more complicated treatment of the concept of charge. In place of the single sort of electric charge which acts as the source of the electromagnetic force, there are *three* sorts of 'charge' for the strong force. Known as *colour*, these sources have been arbitrarily dubbed red, green and blue.

The last of the four forces is known as the weak force. It acts on all quarks and all leptons, with a strength that is far weaker than electromagnetism, but still much stronger than gravity. The weak force manifests itself mainly by bringing about the transmutation of particles rather than exerting a direct pushing or pulling effect. It was intitally postulated to explain beta decay, a type of radioactivity exhibited by some unstable nuclei. In a typical beta decay a neutron turns into a proton, an electron and an antineutrino. This process, which is driven by the weak force, involves a change of quark flavour: within the neutron, one of the up quarks turns into a down. The weak force is capable of changing the flavour of both quarks and leptons. In the latter case, electrons can turn into neutrinos, and so on.

Neutrinos are subject only to the weak force (and gravity of course) so they are exceedingly weakly interacting. Indeed, a neutrino could penetrate many light years of solid lead before being stopped. Nevertheless, the pushing power of neutrinos can actually be observed directly — during the cataclysmic death of stars. Once every few decades per galaxy, a star explodes in an event known as a supernova. Over the centuries many have been witnessed. The most recent occurred in the Large

Magellanic Cloud (a nearby mini-galaxy) in the spring of 1987, and was distinctly visible from Earth.

A supernova is initiated by the sudden collapse of the stellar core under its own weight. During the implosion, an intense pulse of neutrinos is released, and such is the enormous density of stellar material that even these ghostly particles can exert a strong enough force to blast the outer envelope of the star into space, thereby generating an expanding shell of luminous gas. In one of the most exciting observations of the decade, the neutrino pulse from the 1987 supernova was duly detected on Earth a few hours prior to the visual appearance of the stellar explosion.

The range of the weak force is extremely restricted. When the force was first discerned, weak interactions were assumed to be point-like, but it is now known that the range is about 10^{-17} metres.

1.7 Messenger particles

As we saw earlier the behaviour of subatomic particles is governed by quantum theory (see Section 1.4). Any description of forces in the subatomic realm must therefore be consistent with this theory. The starting point of quantum physics was Planck's postulate that light comes in discrete packets or quanta, now known as photons. Electromagnetic disturbances thus propagate through space in the form of photons, which have particle-like qualities. The photon is neither a quark nor a lepton; it instead forms the first member of a third distinct class of particles.

Recall from Section 1.5 that although hadrons can be either fermions or bosons, their constituents the quarks are fermions. Likewise the leptons are fermions. Thus the fundamental particles of matter are all fermions. The photon differs from either quarks or leptons in being a fundamental boson. It has, in fact, a spin of one unit. The mass (strictly rest mass) of the photon is zero; by definition it travels at the speed of light.

The existence of photons must be taken into account when discussing the action of the electromagnetic force. Figure 6 shows the paths of two electrically charged particles according to the classical picture. As the particles approach, the electromagnetic field of A acts on B causing a repulsive force to deflect B away, and vice versa. During this so-called scattering process, momentum and perhaps energy are exchanged between the particles.

In the quantum theory, momentum and energy are 'quantized', i.e. they cannot vary continuously, but are restricted to certain discrete values. As a result, the activity depicted in Figure 7 must be interpreted somewhat differently. Instead of a continuous flow of momentum and energy between the particles through the electromagnetic field, the interaction takes place through the exchange of photons. Figure 7 shows a one-photon exchange, where the wavy line denotes the photon. The direction of passage of the photon cannot be discerned because of the uncertainty principle: the emission and absorption events occur within the interval Δt over which the time is uncertain. Viewed this way, the photon acts as a

Figure 6. In classical physics the mutual deflection of similar electrically charged particles is described as a continuous transfer of momentum causing the particles' trajectories to curve away from each other.

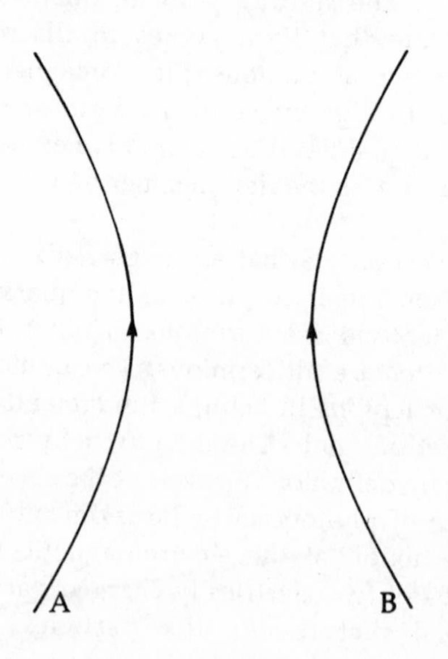

kind of messenger, conveying the electromagnetic force between charged particles. Physicists say that charged particles such as electrons 'couple' to photons, which are thereby responsible for the electromagnetic interaction.

Two-photon exchange also occurs, but it contributes considerably less to the scattering effect than one-photon exchange. Three-, four-,... photon processes are correspondingly feebler. Pictures such as Fig. 6 and 7 are known as Feynman diagrams after Richard Feynman, and the theory behind it is, as mentioned on page 20, called quantum electrodynamics (QED). Detailed calculations of scattering and other electromagnetic processes using these ideas have proved astonishingly successful, and lead to results that are in agreement with measurement to very great accuracy.

It is possible to view all the forces of nature in this way. Each force possesses one or more associated messenger particles. In

Figure 7. The force of repulsion between two charged particles such as electrons may be computed from the effect of the transfer of photons between them.

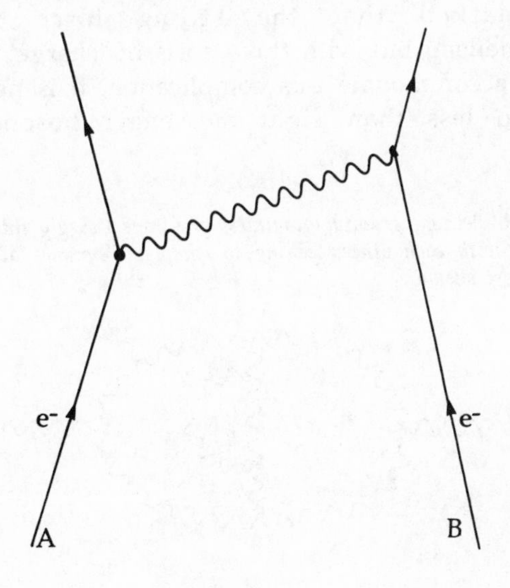

the case of gravity, the analogue of the photon is called the graviton. It too is a massless boson, but with a spin of two. Gravitons couple so weakly to other particles that their effects have yet to be discerned in the laboratory; their existence is accepted on indirect evidence and on grounds of physical consistency. The masslessness of photons and gravitons is directly related to the long-range nature of electromagnetic and gravitational forces.

Another important difference between the graviton and the photon is that whereas photons couple only to charged particles, gravitons couple to all particles, including gravitons! This means that gravitons feel gravity too, and gravitons may interact with each other, leading to processes of the sort shown in Fig. 8. Such tangled webs of gravitons imply that the theory is 'nonlinear', which is to say that graviton processes cannot be simply superimposed on each other. In the case of a linear theory, such as electromagnetism, beams of photons may cross, for example, without mutual disturbance. The inherently nonlinear nature of gravity is the cause of much of the difficulty in providing a quantum description of the force (see Section 1.12).

We remarked that the strong force resembles electromagnetism, but with three sorts of 'charge', known as colour. To accommodate this complication, it is necessary to provide no less than eight messenger bosons. Known

Figure 8. Because gravity gravitates, gravitons (wiggly lines) can interact with each other, leading to complex Feynman diagrams of the sort shown.

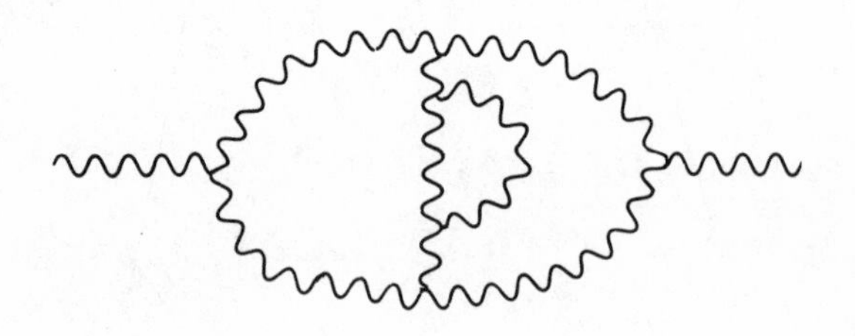

collectively as *gluons*, these quanta of the strong force field also have spin 1, like the photon. However, they share with the graviton the property of coupling to themselves, i.e. gluons as well as quarks are 'coloured'. The resulting theory, known as quantum chromodynamics (or QCD for short) is therefore non-linear. As a result the inter-quark force displays an unusual dependence on distance. Most forces decline with distance, but the gluon force does the opposite. At close range (corresponding to high energies — see page 20) the force fades out, but as the quarks separate from each other the strength grows. In this respect, the gluon force resembles a string of elastic. It seems likely that the force would grow without limit. If that is the case then the quarks will be forever confined within their hadronic prisons.

Finally, the weak force possesses three messenger particles, known as W^+, W^- and Z. All three are spin 1 bosons, but they differ from all the other messenger particles in that they have a nonzero mass. In fact, the masses are very large indeed (about 80 proton masses for the Ws and 90 for Z). It is this fact that is responsible for the very short range of the weak force. The Z is in all respects other than mass like the photon. The Ws, however, are electrically charged; in fact the W^- is the antiparticle of the W^+. Both have one unit of charge, like the electron.

1.8 Symmetry and supersymmetry

A proper treatment of symmetry involves advanced mathematics and is beyond the scope of this book, but the basic concepts are not hard to grasp. To fix ideas, consider some simple geometrical line shapes: square, equilateral triangle, circle (see Figure 9). Each has rich and interesting symmetry properties. Perhaps the most familiar, possessed by all three, is reflection symmetry. If a mirror is placed perpendicular to the page along the broken lines (try this), the shapes remain unchanged. In each case, the left-hand side of the figure is the reflection of the right-hand side. A fancy way of expressing this is to say that the figures remain invariant under reflections in the broken line axes. Notice that several axes of reflection symmetry exist for each figure: four for the square, three for the triangle and an infinite number for the circle (placing the mirror along any diameter will do).

We can find other symmetries in these figures. If the triangle is rotated through 120°, 240° and 360° about the dot in the middle, it looks identical. The square can be rotated to four positions 90°, 180°, 270°, 360° and remain the same. One says that they are invariant under rotations through multiples of 120° and 90°, respectively. In the case of the circle any rotation whatever about its centre leaves the circle invariant. Symmetry, therefore, comes in two distinct varieties: continuous and discrete. The rotation of the circle is a continuous operation, at all times leaving the shape unchanged. On the other hand, the rotations of the square and triangle, or the reflection symmetries, are discrete.

An interesting perspective on symmetry is obtained if we ask what it is that really characterizes the superior symmetry of the circle over that of, say, the square. One way of looking at this is to see that the square has more structure than the circle. Compared to a square, a circle is rather featureless. We could destroy the rotation symmetry of the circle by flattening it a bit, or painting a dot on it. In both cases the result is to add new features and structure. As a general rule systems with few features possess more symmetries.

Perhaps the most extreme example of a featureless system is empty space. Nothing changes if we imagine it to be rotated. It would also remain the same if it were displaced (i.e. translated)

Figure 9. Examples of geometrical symmetries. Each figure shown is unchanged if reflected in one of the broken lines.

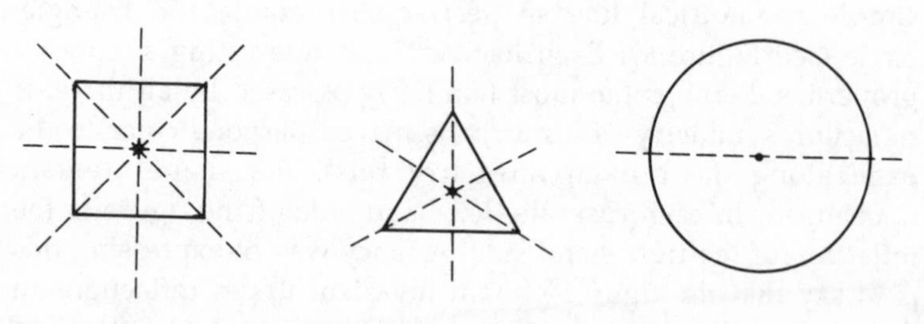

in any given direction. We might say that empty space is the same in all directions and at all positions. (This is only true to the extent that we ignore curvature effects associated with gravity. This is almost always justified in particle physics.) In addition, empty space is unchanged by mirror reflection. These elementary observations can be made more precise by saying that the geometrical structure of space, i.e. distances and angles, is invariant under continuous rotations and translations, and under reflections in any plane.

In this featureless, empty world, time also possesses symmetry. In a void in which nothing happens, one moment of time is as good as any other. This means that there is invariance under time translations, too. There is also invariance under temporal reflections, i.e. time reversal. Nothing happens in an empty world, so there can be no distinction between past and future time directions.

In the real world, of course, space is not totally empty. It abounds with fields and particles and ceaseless quantum activity. The symmetries which are exact for empty space are broken, although there may still exist approximate symmetries. For example, in the solar system not all directions are equivalent: towards the Sun things obviously look very different from directions away from the Sun. However, for many purposes the departures from exact symmetry are unimportant and can be ignored to good approximation.

To be specific, consider an isolated particle residing somewhere in outer space. The particle could be either a billiard ball or an atom (we will ignore quantum effects for the time being), and we shall suppose that bodies such as the Sun, as well as other particles, are too far away to have much influence on the behaviour of the particle, and the effects of any force fields are negligible. It would be surprising if the particle were suddenly to fly off in a particular direction. We should suppose that some external force had been overlooked. In the absence of any force, we are sure that the particle would not move. The basis for this confidence is precisely our unquestioning assumption that space is symmetric under translations. If one part of space is the same as every other, why should one place be distinguished by the sudden arrival of a particle? Moreover, why should the particle choose one particular direction in which to fly off, rather than any other?

Similar reasoning can be applied to rotations. We should not expect a body suddenly to start spinning without external propulsion, for why should it spin, say, clockwise rather than anticlockwise? Furthermore, a body spins about an axis, which defines a special direction in space. If space is symmetric under rotations, no direction is special. Hence we do not expect a body to start spinning spontaneously.

These crude observations can be made mathematically precise and provide a deep and powerful connection between, on the one hand, the geometrical symmetries of space, and on the other, the dynamical behaviour of material bodies. In fact, forbidding the absence of spontaneous changes in motion amounts to a statement of the laws of conservation of momentum and angular momentum. The translation symmetry of space leads directly to momentum conservation for particles, whereas the rotational symmetry implies angular momentum conservation. In addition to this, the conservation of energy can be shown to follow from the translation symmetry of time (one moment is as good as any other). Thus, the most fundamental and comprehensive laws of physics are seen to follow from the elementary and unsurprising fact that empty space and time are featureless. It well illustrates the power of symmetry in ordering the natural world.

An interesting question now arises. Do all the forces of nature automatically respect the geometrical symmetries of space and time? Maxwell's electromagnetic theory certainly incorporates all the symmetries we have just discussed, as does our best description of gravity. For a long while, physicists assumed that the nuclear forces must also enjoy the full range of geometrical symmetries. It would, of course, be alarming if the conservation laws of energy, momentum and angular momentum were to fail in the subnuclear world.

What about the discrete geometrical symmetries? How can the laws of physics be tested for them? One way to approach this is to ask: suppose someone shoots a movie film of some particular natural activity and then projects it onto a mirror (or alternatively with the film back to front), would we notice the deception? Would we see in the mirror any apparently impossible process? Likewise, if the film is run backwards, would we witness any events that seem to violate the laws of physics?

To take a simple example, suppose the film shows a spinning sphere (see Figure 10). The spin axis defines a particular direction and we can draw a line along it. If we watch the spinning sphere in a mirror its 'handedness' is reversed — clockwise and anticlockwise spins become interchanged. We would thus see a sphere spinning the opposite way. But there is nothing special about the direction of spin. So provided we were not aware of the mirror we would have no reason for suspecting any trickery. Of course, if on closer inspection the sphere turned out to be the Earth, the deception would be obvious because dawn would seem to come across the continents from west to east instead of east to west. However, we are supposed to be discussing the symmetry of the laws of physics, not the symmetry of particular large objects in the real world. In the world of subatomic particles, there are no 'continents' to distinguish one particle from another, so such incidental complications do not arise.

The example of the rotating sphere also serves to illustrate time-reversal symmetry, for a movie film played backwards will also show a reversal of the spin. From the images alone we would not be able to tell whether the film was being played backwards or forwards — both look equally plausible if the sphere is featureless. Again, in everyday life we all know how easy it is to spot something wrong when the same thing is done with shots of a towerblock under demolition, a man painting a house, or perhaps a bucket of water being spilt. But in the microworld there is nothing remarkable about the reversal of spin. The same applies to other familiar processes, such as the collision and disintegration of particles — their time reversals do not appear miraculous either. Only when the activities of many particles together are reversed do we suspect something. For example, the spontaneous disintegration of a neutron into proton, electron and antineutrino if shown in reverse would invite scepticism because we would see an exceedingly improbable triple encounter of proton, electron and antineutrino. In the case of most macroscopic processes, the odds against their reversal occurring are astronomical.

Symmetry in daily life is most obvious in geometry (see Figure 9) though it may occur in many other ways. Symmetry in time is one example already discussed. There are other important symmetries in physics not directly connected with space or time, and these prove to be of the greatest importance.

Figure 10. Reflection symmetry. The spinning sphere rotates clockwise in the real world and anticlockwise in the mirror world. The latter is unexceptionable, and if we didn't see the edges of the mirror we could not tell which sphere was real and which was an image. Both are equally possible situations. The mirror view would also appear if a movie film of the spinning sphere was played backwards.

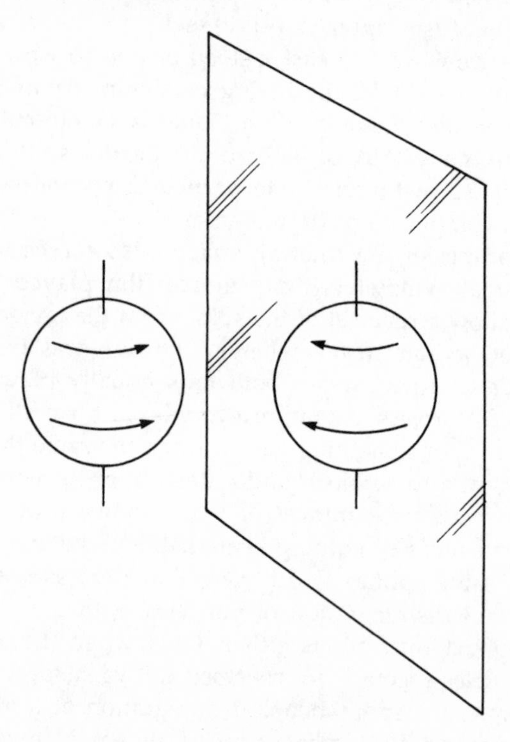

One simple case is the discrete symmetry of electric charge reversal. The electron and the positron have already been described as 'mirror' particles, and in a sense we can think of the positron as the charge 'reflection' of the electron. The fact that the magnitude of the charge is the same for both means that they are exact reflections, so we might expect the laws of physics to be invariant under charge reflection too.

There is an interesting mathematical theorem which proves that, subject only to some very weak assumptions which nobody seriously doubts, the laws of physics must be invariant under the combined operations of space reflection, time reversal and charge reflection. The operations are usually denoted by P, T and C respectively (P is used for space reflection because this operation is also known as 'parity' change). The theorem is known as the CPT theorem.

In the mid-1950s physicists had difficulty making sense of some processes in which hadrons decayed by the weak force. Two Chinese Americans, Tsung Dao Lee and Chen Ning Yang made the bold suggestion that perhaps the law of parity conservation is violated by the weak force. Although everyone had assumed that parity would always be conserved, nobody had actually tested this specifically for the weak force. Another Chinese American, Chien-Shiung Wu, then performed a classic experiment to determine the reflection properties of a weak interaction process.

Her experiments consisted of examining the direction of emission of beta particles from the nuclei of cobalt 60. The objective was to determine the direction relative to the spin axis of the cobalt nuclei. The situation is depicted in Figure 11, with the convention that the spin vector points along the spin axis in the direction of the advance of a right-handed corkscrew. Wu found that the electrons preferred to come out in a direction pointing away from the head of the spin vector. Viewed in a mirror, this preference is reversed, so the decay clearly has a preferred 'handedness'. That is, if a movie film of the experiment were projected onto a screen from a mirror, a physicist would spot that the image was reflected: the mirror image shows an impossible process. The experiment therefore demonstrated directly that parity is not conserved in beta decay.

The nonconservation of parity is found to be a general feature of weak interactions. A clear demonstration is provided by the decay of the negatively charged muon (denoted μ^-) into an electron (e^-) plus neutrinos. Although the fate of the neutrinos cannot be directly monitored, it is possible to measure both the direction of the muon's spin and the direction of motion of the emitted electron when the muon decays. It is found that although the electron can be emitted from μ^- over a whole range of angles relative to the muon's spin axis there is a preference for the electron to fly off towards the side from which the spin of the muon appears clockwise, rather than the other side.

The situation is depicted in Figure 12. Look at the mirror image of the decay. When reflected in a mirror the spin of the muon is reversed from left-handed to right-handed rotation. In the mirror image the electron chooses to emerge on the side from which the muon appears to be spinning anticlockwise. The mirror therefore alters the relationship between the spin

Figure 11. Parity violation. Wu's experiment demonstrated that beta particles emitted by decaying nuclei of Co^{60} prefer to emerge away from the direction of the nuclear spin vector. Viewed in a mirror this asymmetry is reversed.

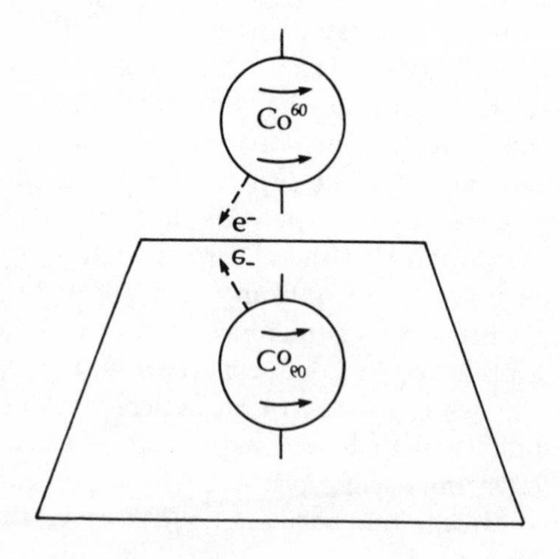

direction and the direction of the emergent particle. If a similar analysis is carried out for the antiparticles, in which a positively charged muon (μ^+) decays into a positron (e^+), it is found that the situation is reversed, and appears like the arrangement shown in the mirror image in Figure 12. Moreover this reversal is exact – there is precisely the same degree of lopsidedness, but in the opposite direction. This result accords with the symmetry between matter and antimatter. It implies that, under the combined operation of parity reversal P and charge reversal C (which changes μ^- to μ^+) the laws governing muon decay are invariant. Thus, even if P is violated, CP remains a symmetry.

Figure 12. When μ^- decays, the electron prefers to emerge towards the right side of the spin axis as shown, rather than the left (two neutrinos, not shown here, are also emitted). Clearly this lopsided tendency is asymmetric under reflections: the mirror image shows the electron choosing the left side. This is the behaviour adopted by the antiparticle decay.

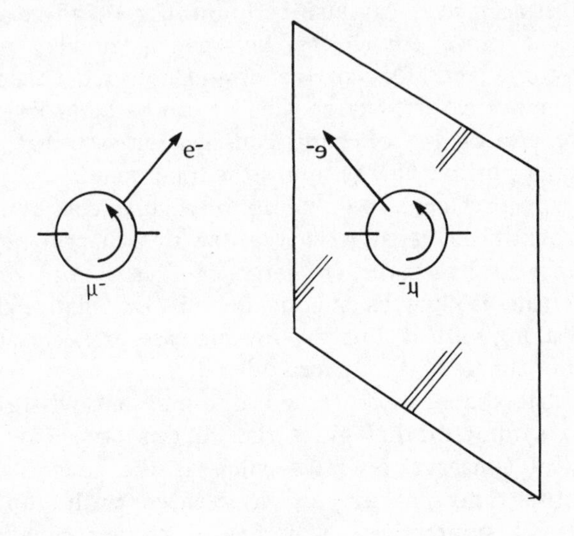

The discovery of parity violation (i.e. nonconservation of parity) in weak force processes came as a great shock to physicists. Although the world abounds with complex structures that possess a definite handedness (e.g. DNA), the existence of a preferred handedness in the *laws* of physics is altogether more profound. It implies that, even in the absence of complex structures, nature distinguishes left-handed from right-handed. Hitherto it was assumed that nature would no more distinguish left from right than it would distinguish up from down in empty space. The history of physics shows that formidable advances can be made through mathematical exploration, particularly when the concept of symmetry is exploited. Even though the mathematical symmetries may be hard, or even impossible, to visualize physically, they can point to important new principles of nature. Searching for undiscovered symmetries has become a powerful method in helping physicists advance their understanding of the world.

So far, the continuous symmetries that we have discussed all involve space, or spacetime. But continuous symmetries of a more abstract nature can also be found. As already explained, there is a close connection between symmetry and the conservation laws. One of the best-established conservation laws is that of electric charge. Charge can be both positive and negative, and the law of charge conservation says that the total quantity of positive charge minus the total quantity of negative charge cannot change. If a quantity of positive charge meets an equal quantity of negative charge, the two charges neutralize each other to give zero net charge. Similarly, positive charge can be created so long as an equal quantity of negative charge is created along with it. But any net increase or decrease in the amount of charge is strictly forbidden.

If electric charge is conserved, one may ask whether there exists a symmetry that gives rise to this law. The various dynamical conservation laws, such as the conservation of momentum and energy, are associated with continuous geometrical symmetries. The law of charge conservation, however, refers to an abstract rather than a dynamical property, which suggests that an abstract symmetry is responsible. As an example of such an abstract symmetry taken from everyday life, consider the phenomenon of economic inflation. As the real value of the pound or dollar declines, so the wealth of someone on a fixed income declines with it. If, however, a person has an index-linked income, their earning power is independent of the value of the currency. A formal way of putting that, but one which will be useful for reasons that will be apparent in a moment, is to say that index-linked income is symmetric under inflationary changes.

In physics there are many examples of non-geometrical symmetries. One of these concerns the work necessary to lift a weight. The energy expended depends on the difference in height by which the weight is raised (it does not depend on the route taken). The energy is independent, however, of the absolute height: it would not matter whether the heights were measured from sea level or ground level, because it is only the height difference which is involved. There is a symmetry, therefore, under changes in the choice of zero height.

A similar symmetry exists for electric fields, for which voltage (electric potential) plays a role analogous to height. If an electric charge is moved from one point to another in an electric field, the energy expended depends only on the voltage difference between the end points of its path. If the voltage is raised by the same amount at both ends of the path, the energy expended does not alter. This is therefore an important symmetry of Maxwell's electromagnetic equations.

The three examples given above illustrate what physicists call gauge symmetries. One can think of the symmetries involved as a 'regauging' of money, height, and voltage respectively. All are abstract symmetries, in the sense that they are not geometrical in nature. We cannot look at them and see the symmetry. Nevertheless, they are still important indicators of the properties of the system concerned. Indeed, it is precisely the gauge symmetry for voltages that ensures the conservation of electric charge. Gauge symmetries have played a central role in the search for a successful quantum theory of the various forces of nature, and it is within the context of gauge symmetries that attempts to unify the forces have been made.

We have seen how symmetries in physics divide into geometrical symmetries, such as rotations and reflections, and abstract symmetries, such as gauge symmetries. In the early 1970s theorists unexpectedly discovered a hitherto unknown sort of geometrical symmetry which is deeper and more powerful than such familiar operations as rotations and translations. It is called supersymmetry.

In Section 1.5 it was mentioned that the geometrical structure of space experienced by a fermion differs fundamentally from that experienced by a boson: a fermion has to be rotated through 720° before it returns to its starting configuration. This 'double-valued' quality of fermions implies that the algebra of geometrical symmetry operations such as rotations is radically different for fermions than it is for bosons, and ordinary objects. Indeed, one of the reasons why the distinction between bosons and fermions is so fundamental is precisely because they possess completely different geometrical properties.

The novel feature of supersymmetry is that it provides a geometrical framework within which fermions and bosons receive a common description. This cannot be achieved by staying within the context of familiar geometrical operations in ordinary space. It is possible to represent supersymmetry operations mathematically by attaching to the four dimensions of ordinary spacetime another four dimensions, forming something known as 'superspace'. The purpose of the extra four dimensions is to accommodate the peculiar geometrical properties of fermions, so these additional 'fermionic dimensions' are not space or time dimensions as we know them.

The rules of geometry in the extra dimensions are very strange. As an example of the difference, consider the rotation operation. It is easy to verify that if two rotations are performed in succession, then the outcome of the combined operation depends on the order in which the two are performed. Figure 13 illustrates this in the case of 90° rotations of a book. The final orientation of the book is completely different in the two cases. If the successive rotation operations are denoted by R_1 and R_2, this difference can be expressed symbolically by writing $R_1R_2 \neq R_2R_1$, or equivalently $R_1R_2 - R_2R_1 \neq 0$. The combination of $R_1R_2 - R_2R_1$ is known as the *commutator* of R_1 and R_2. Starting with such 'commutator' relations one can

Figure 13. Rotations do not commute. In the top sequence, successive 90 degree rotations are applied to a book, about a vertical and horizontal axis respectively. In the lower sequence the order of rotations is reversed. The end result of the two sequences is different.

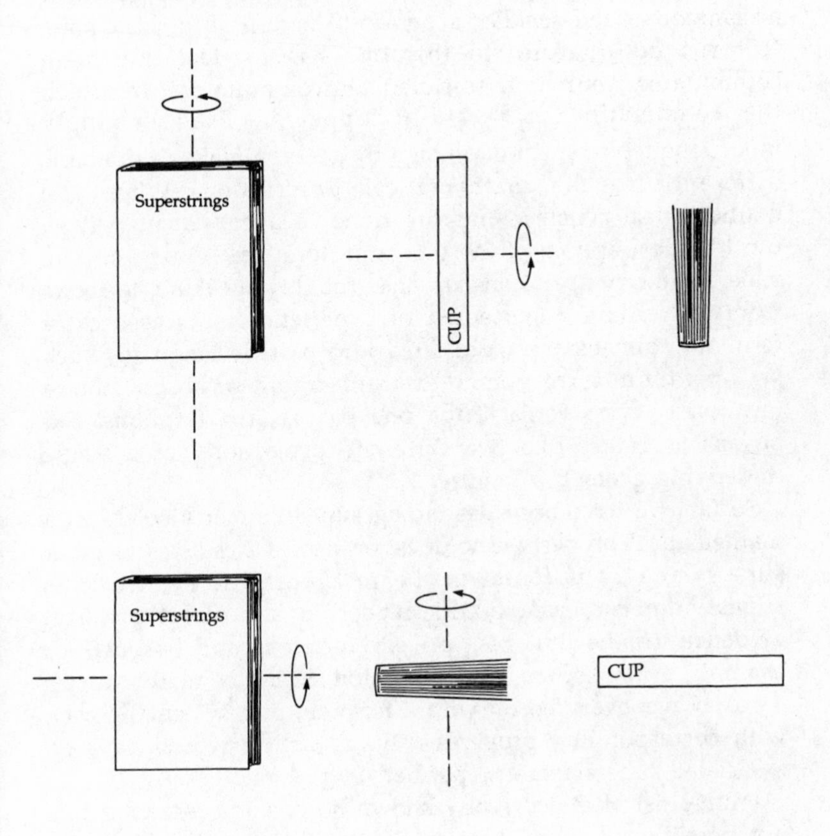

construct an algebra of rotations, which embodies the geometrical properties of space as experienced by books and bosons. It is possible to define the analogue of rotations R in the extra four fermionic dimensions of superspace too. However, the geometrical properties of this part of superspace must accommodate the peculiar geometrical nature of intrinsic spin. It turns out that to do this one has to deal, not with commutators, but with so-called anticommutators, in which the combination $R_1R_2 + R_2R_1$ appears. The seemingly innocuous replacement of a - sign by a + sign makes a dramatic difference to the mathematical description. When the mathematical constructions are done in a consistent way, a unified description of bosons and fermions emerges. The supersymmetry operations are able, roughly speaking, to rotate from the ordinary spacetime of experience into these extra fermionic dimensions. Translated into particle language, such an operation corresponds to tranforming a boson into a fermion or vice versa. Thus one can regard fermions and bosons as rather like two different 'projections' of a single underlying geometrical entity.

So far, we have been describing supersymmetry solely as a mathematical property. The question now arises as to whether supersymmetry is found in the real world. If the world is supersymmetric, we should expect to see direct physical evidence for the linkages between fermions and bosons. For example, every known type of fermion should be matched with a boson and every boson with a fermion, in a systematic way, with corresponding properties. Each particle would have an associated supersymmetric partner.

From a list of the currently known bosons and fermions, it is not possible to marry them up in this fashion. This does not necessarily mean that supersymmetry is irrelevant to the real world, however. First, it is often the case in nature that a deep symmetry of the laws of physics is actually broken in the physical state of the system. Such symmetry breaking occurs, for example, in the so-called electroweak force (see p.56), where the underlying symmetry of the force is hidden. It could be the case that nature is basically supersymmetric, but that the symmetry is broken in most phenomena so far investigated.

Secondly, there is no reason why the currently known fermions have to be superpartners for the known bosons. There may be many undiscovered particles that are the superpartners to the known particles. Thus it has been postulated that there exists, for example, so-called photinos, these being the superpartners of the photons. The reason, it is said, why nobody has yet detected a photino is because they interact only very weakly with ordinary matter and so would not have been readily observed in a detector. Similarly, supertheorists talk about gluinos to go with the gluons and gravitinos to go with the gravitons. Then there are the bosonic superpartners for the fermions, dubbed squarks and sleptons. At present all these exotic superpartners are conjectural. Supersymmetry is therefore a powerful theoretical idea, but one which lacks any unequivocal confirmation in nature.

1.9 Unification of the forces

When Michael Faraday discovered the phenomenon of electromagnetic induction in the 1830s he established a clear link between two forces of nature – electricity and magnetism – although it took Maxwell's mathematical insight to provide a fully unified electromagnetic theory in the 1850s. Nevertheless in 1850 Faraday had already surmised that there may be a further link between electricity and gravitation. To investigate this idea, Faraday constructed several ingenious pieces of apparatus which would serve to show whether, for example, falling bodies generate electric fields. A cartoon illustration is shown in Figure 14. The results of Faraday's experiments were negative, but this did not dispel his faith that, deep down, electric and gravitational forces are part of a common superforce.

The next serious attempt at a unified gravitational–electromagnetic scheme came in 1921. This was just a few years after Einstein had published his gravitational theory – the general theory of relativity. As explained in Section 1.3, an important feature of this theory is that space and time are joined into a four-dimensional spacetime. Contemplating this, a German mathematician, Theodor Kaluza, decided to write

Figure 14. Cartoon of Faraday's attempt to demonstrate a link between electric and gravitational forces (cartoon courtesy of A. de Rujula).

down Einstein's gravitational field equations in five dimensions instead of four, by the simple expedient of appending an imaginary additional dimension of space. The result was unexpectedly fruitful. Viewed in the usual context of four dimensions, the five-dimensional gravitational field equations of Einstein reproduce the usual four-dimensional gravitational equations *plus* another set of equations that turn out to be precisely Maxwell's equations for the electromagnetic field. Thus, by formulating gravity in five dimensions one can obtain *both* gravity and electromagnetism from a single theory. In other words, according to Kaluza's theory, electromagnetism is not a separate force, but an aspect of gravity, albeit in a world involving an unseen higher dimension of space.

The principal weakness of the theory is the fact that we perceive only four dimensions in the real world. If the idea of five dimensions is to be taken seriously it is necessary to explain what has become of the fifth dimension. In 1926 the Swedish physicist Oscar Klein came up with a marvellously simple answer. Klein proposed that we do not notice the extra dimension because it is, in a sense, 'rolled up' to a very small size. The situation can be compared to a hosepipe. Viewed from afar, the pipe appears to be just a wiggly line. On closer inspection, what we took to be a point on the line turns out to be a circle around the circumference of the tube (Figure 15). Suppose, conjectured Klein, that our universe is like that. What we normally think of as a point in three-dimensional space is in reality a tiny circle going round the fourth spatial dimension. From every point in space a little loop goes off in a direction that is not up, down, or sideways, or anywhere else in the space of our senses. The reason we have not noticed all these loops, so the argument goes, is because they are incredibly small in circumference.

Klein's idea takes a bit of getting used to. Part of the trouble is that it is hard to visualize where those loops are looping. The loops are not *inside* space, they extend space, just as a wiggly line moved rigidly around a loop marks out a tube. We can readily envisage the situation in two dimensions, but not in four. Nevetheless, the proposition still makes sense. Klein was able to calculate the circumference of the loops around the fifth dimension from the known values of the unit of electric charge carried by electrons and other particles, and the strength of the gravitational forces between particles. The value came out to be 10^{-30} centimetres, or about 10^{-17} of the size of an atomic nucleus. It is no surprise that we have not noticed this putative fifth dimension, for it would be rolled up to a size far smaller than any structure yet discerned, even in subnuclear particle physics.

In spite of its ingenuity, the Kaluza–Klein theory remained little more than a mathematical curiosity for over fifty years. With the discovery of the weak and strong forces in the 1930s, the idea of unifying gravity and electromagnetism lost much of its appeal. Any successful unified field theory would have to accommodate not just two, but four forces. This step could not be accomplished, therefore, until scientists had attained a proper understanding of the weak and strong forces.

Figure 15. From a distance a hosepipe looks like a wiggly line. On closer inspection a point P on the line turns out to be a circle around the circumference of the pipe. It is possible that what we normally regard as a point in three-dimensional space is really a tiny circle around another dimension of space. This idea forms the basis of Kaluza and Klein's unified theory of electromagnetic and gravitational forces.

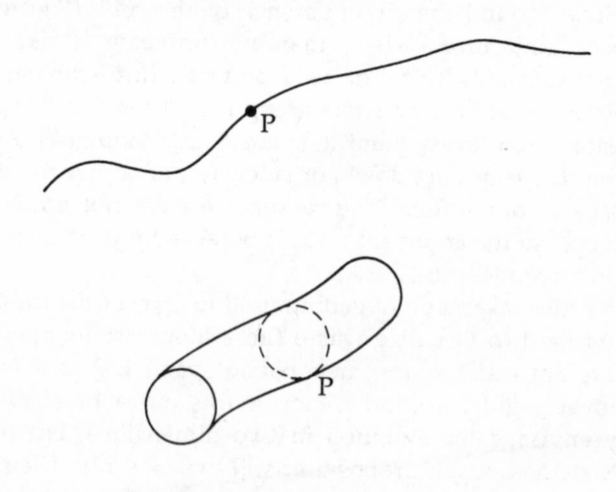

The burgeoning study of subnuclear particles and their forces that took place in the 1950s revealed a picture of bewildering complexity that forestalled any hopes for a simple sort of unification. Indeed, of the four known forces, only electromagnetism could lay claim to a theory (QED, see page 31) that was both internally consistent, and consistent with the all-important theories of relativity and quantum mechanics.

The other three forces were not at all well understood at that time. The discovery of parity nonconservation required a complete overhaul of the theory of the weak force to incorporate the left–right asymmetry into its action. This was done, but the resulting theory gave sensible answers only for some simple weak force processes, and then only so long as the energies involved were not too high. For most processes the answers were nonsensical. The theory was thus mathematically inconsistent, had little predictive power, and was clearly fundamentally flawed.

The strong force was not understood at all. Indeed, the interaction between hadrons seemed to involve a complex tangle of different forces and fields rather than a single strong nuclear force. Today, we know that the inter-hadron force is really just a complicated vestige of the more primitive inter-quark force, but in the early days attempts were made to treat the force between hadrons as fundamental. As early as 1935 the Japanese physicist Hideki Yukawa modelled the strong force by analogy with QED, invoking the exchange of a strong force 'messenger' particle between hadrons, so leading to the successful prediction of the pion. In spite of this, it soon became clear that the pion exchange model gave only a crude description of the strong nuclear force. Furthermore, as in the case of the weak force, calculations for strong force processes often yielded nonsensical answers.

Gravity had a peculiar status in the 1950s. Unlike the weak and strong forces, gravity possessed a consistent and very elegant theoretical formulation at the classical (i.e. nonquantum) level, namely, Einstein's general theory of relativity. Moreover, it led to some distinctive predictions that had been confirmed by experiment. The main difficulty with general relativity came when physicists attempted to provide a quantum mechanical description of gravity. As in the theory of the weak force, mathematical inconsistencies rendered the theory impotent for predicting all but the simplest processes.

The difficulties with the quantum formulation of gravitation were ignored by the vast majority of physicists in the fifties and sixties, because gravity is usually only conspicuous on astronomical scales where Einstein's classical theory is perfectly adequate. Gravitons couple far too weakly to other particles for them to be observed or to play a direct role in particle physics. However, the difficulties with quantum gravity were if anything rather more serious than those of the weak and strong forces. The general theory of relativity occupies a central place in twentieth century physics, and not only on account of its predictive successes. The theory is founded upon very deep, clear and elegant principles, it is fundamentally simple and mathematically attractive, and it reduces gravity to pure geometry. It therefore has compelling aesthetic and philosophical appeal.

The quantum theory has a rather different status. It lacks the inherent simplicity and aesthetic appeal of general relativity. Its founding postulates are counterintuitive, and there are grave doubts about its philosophical consistency in relation to the observer. (For a survey of this last point, the reader is referred to our book *The Ghost in the Atom.*) On the other hand its successful applications far exceed those of general relativity. Quantum mechanics is an indispensible part of particle physics, nuclear, atomic, molecular and solid state physics, physical chemistry, modern optics, stellar astrophysics and cosmology.

It is often said that twentieth century physics is founded upon the theory of relativity and the quantum theory. The first of these is extremely beautiful and convincing, but has limited application, the second is somewhat messy but has a proven track record without equal in science. The fact that these two theories are *incompatible* implies a profound and devastating incosistency in the heart of physics. Any successful TOE must somehow remove this inconsistency.

1.10 Unified gauge theories

For many years the problems of quantum gravity were considered to be completely intractable. The issue was largely put to one side while physicists turned their attention to the weak and strong forces. In the early sixties Sheldon Glashow and others spotted that although the weak and electromagnetic forces are superficially rather dissimilar, at a deeper level they share a number of common features. Both, for example, are conveyed by exchanging spin 1 bosons. Furthermore, it is possible to understand the weak force in terms of a weak 'charge' and a weak 'current' in many ways analogous to the concepts of electric charge and current.

The principle difference between the two forces is the fact that the photon is massless and the electromagnetic force is long-ranged, whereas the quanta of the weak force field are very massive and the force is very short-ranged. If the weak force shared the long-range nature of the electromagnetic force, the two would be almost identical. Physicists began to appreciate that it might be possible to write down an amalgamation of the two forces in a common theory, thus extending the programme of unification initiated by Maxwell in the nineteenth century.

A mathematical analysis reveals that the masslessness of the photon is intimately connected with the gauge symmetry that Maxwell had incorporated into the electromagnetic field equations. It is this gauge symmetry that turns out to be the all-important property which ensures the mathematical self-consistency of QED. By contrast, the massive nature of the weak force messenger particles breaks any gauge symmmetry that might be present in the underlying dynamics. It was this broken gauge symmetry that made the early theories of the weak force unsatisfactory. Circumventing this difficulty offered the hope of finding a consistent theory of the weak force, and also of unifying the weak force with the electromagnetic force.

In the late 1960s, Steven Weinberg and Abdus Salam independently realized that it might be possible for the weak messenger particles to acquire a mass without breaking the underlying gauge symmetry of the weak force. Rather than build a mass into the theory at the fundamental level – in the dynamical equations themselves – this mass could arise 'spontaneously' as a result of certain interactions that take place in the weak force field. The appearance of a mass could thus be made a secondary affair, leaving the gauge symmetry in the dynamics itself intact.

The idea that the weak gauge symmetry might be spontaneously rather than dynamically broken is based on analogy with other forms of spontaneous symmetry breaking that are familiar in many branches of physics. A simple example from elementary classical mechanics is illustrated in Figure 16. Imagine a ball at the top of the 'Mexican hat' surface shown in the diagram. In this position the state of the system is clearly symmetrical under rotations about the vertical axis through the top of the 'hat'. Moreover because gravity acts vertically, there is no preferred horizontal direction to the system: the forces acting are therefore also rotationally symmetrical. In this configuration, then, the position of the ball (i.e. the state of the system) reflects the underlying symmetry of the forces acting. However, this state is obviously unstable. If the ball is released it will roll down the side of the surface and, by dissipating its energy, eventually come to rest somewhere in the 'rim of the hat' (Figure 16). The latter configuration is stable, but the rotational symmetry has been broken. The actual position in the rim chosen by the ball has no deep significance:

Figure 16. Spontaneous symmetry breaking. The ball is placed on top of the 'Mexican hat' surface. In this configuration there is complete rotational symmetry. However, the configuration is unstable, and the ball spontaneously rolls down into the rim of the 'hat', coming to rest at some arbitrary point. The rotational symmetry is thereby broken. The system has traded symmetry for stability.

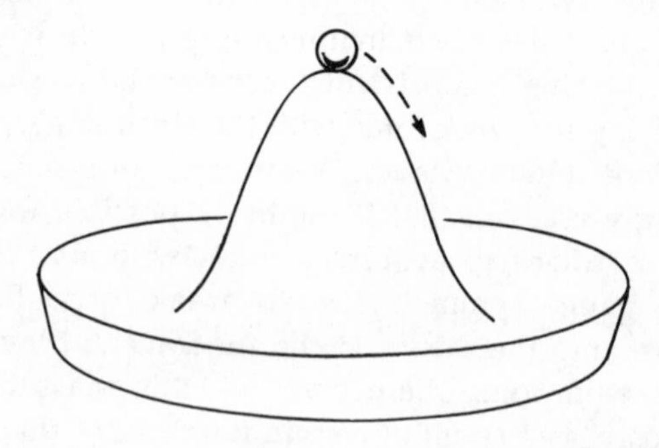

it is purely random. Yet by selecting a particular orientation relative to the surface, the ball ensures that the state of the system no longer reflects the symmetry of the underlying forces. This type of symmetry breaking, in which the symmetry is still there in the underlying forces but is masked by the lopsided nature of the state, is called 'spontaneous'.

Weinberg and Salam proposed that the W and Z particles acquire their masses through a spontaneous breaking of the underlying gauge symmetry. In this way the crucial symmetry would still be there, but hidden. Given this interpretation of the masses, the weak force could be placed on the same footing as the electromagnetic force, and the two could be given a common description. But in the actual quantum state of the system, W and Z would not reflect the underlying gauge symmetry because of their large masses, although things could be arranged so that the photon still displayed gauge symmetry by remaining massless.

To achieve these results, Weinberg and Salam introduced an additional quantum field, called a Higgs field after its originator, Peter Higgs. The quanta of the Higgs field are massive spinless bosons. The effect of the coupling between the Higgs field and the electromagnetic–weak fields is to introduce a potential energy of precisely the 'Mexican hat' form shown in Figure 16 (though the surface is in an abstract space, not real space as drawn). Under the influence of this coupling the system seeks out the lowest energy quantum state (ball in the rim) which here corresponds to W and Z acquiring large masses.

The Glashow–Salam–Weinberg theory nicely explains the differences in strength between the effective low-energy weak force and the electromagnetic force. The underlying interaction strength of both forces is comparable, and a weak charge g can be introduced analogous to the electric charge e, leading to an *effective* weak coupling constant g/M, where M is the mass of the W particle. Because M is so large (about 80 proton masses) the effective strength of the weak coupling is, as its name implies, extremely small.

The ratio e/g is a free parameter in the theory. Usually it is expressed as an angle θ through the relation $e = g \sin\theta$. The value of θ must be fixed by experiment. It is found to be about

28°. Thus θ determines the relative actual strengths of the two forces.

A key theoretical test of the theory was the subsequent demonstration that the mathematical inconsistencies that bedevilled the old theory of the weak force are absent. Moreover, the high energy behaviour of the new theory is entirely satisfactory. In fact, as the energy of the processes of interest is raised so the differences between the weak and electromagnetic forces declines, and at energies comparable to M (80 GeV where 1 GeV = 10^9 electron volts) the two interactions essentially merge in identity.

As for experiment, the new theory predicts a number of subtle but measurable physical effects. One of these is the scattering, without change in identity, of neutrinos from neutrons, a process that is impossible according to the old theory of the weak force. In 1973 an experiment at CERN involving an intense beam of neutrinos traversing a bubble chamber gave unmistakable evidence that neutrinos were scattering from neutrons in the nuclei of atoms inside the chamber. The most dramatic confirmation of the Glashow–Weinberg–Salam theory came, however, in late 1983 and early 1984, when the W and Z particles were produced for the first time in the high energy proton–antiproton collider at CERN. Their masses conform well to the prediction of the theory using the known value of θ.

These encouraging successes have led to the belief that the weak and electromagnetic forces are really two aspects of a unified electroweak force. However, the parameter θ remains undetermined by the theory, so perhaps 'amalgamated' is a better word than unified. The crucial element in this success was the formulation of the theory in terms of gauge symmetries, and this has encouraged the theoretical examination of a variety of other gauge theories for the description of the strong and gravitational forces, and their eventual unification with the electroweak force.

Gauge symmetries are described using a branch of mathematics known as group theory. A group is a set of mathematical objects (often in practice represented by matrices) that can be combined together by multiplication (subject to certain technical restrictions). Each symmetry can be labelled by the name of the group that generates it. A very simple example concerns the symmetry of the circle. A circle remains symmetric if it is rotated through any angle about its centre. The algebra of such rotations forms a group known as $U(1)$, the U standing for 'unitary', a certain technical mathematical property. It so happens that the gauge symmetry of the electromagnetic field is precisely this $U(1)$ symmetry, but in an abstract space rather than real space.

The electroweak force combines the $U(1)$ group with a slightly more complicated group known as $SU(2)$, the S standing for 'special'. Its detailed properties need not concern us. As discussed in Section 1.6, the strong force has a compelling theoretical description in terms of QCD. This is a gauge theory too, based on a yet more complicated gauge group – $SU(3)$. In the mid-seventies, many attempts were made to unify the electroweak force with QCD to produce a 'grand unified force'. These grand unified theories, or GUTs, are based on finding a larger, more embracing, gauge group, which embodies the $SU(3)$ of QCD and the $SU(2)$ and $U(1)$ gauge groups of the weak and electromagnetic forces, as subgroups. In these schemes θ is no longer a free parameter but is determined by the way in which the large gauge group decomposes into the required subgroups.

A general feature of GUTs is that they mix together the identities of the sources of the three forces. Thus leptons, which give rise to the electroweak force, become associated with quarks which are the source of the strong force. A hint of such an association comes from the fact that there are an equal number of quarks and leptons (at least those currently known). The mixing occurs by the exchange of a new set of messenger particles, collectively known simply as X particles. The exchange of an X particle can turn a quark into a lepton or vice versa.

Once again, at low energies the forces have distinct identities, but at high energies they merge into a single interaction. The energy at which this convergence occurs can be determined from the fact that the inter-quark force rises with separation between the quarks. Recall that the Heisenberg uncertainty principle associates an energy and momentum with a distance and time. Thus low energy experiments reveal the large-separation behaviour of quarks, while high energy experiments probe the short-distance behaviour where quarks approach each other closely. It is possible to compute at what distance – hence energy – the inter-quark force would fall to the strength of the electroweak force. This is the energy at which one would expect unification to be manifest, as all three forces would be of comparable strength. The relevant grand unification energy is about thirteen powers of ten greater than the electroweak unification scale, and far beyond the limits at which experimental tests are feasible.

Fortunately the GUTs make some low energy predictions too. As previously remarked, the theory mixes leptons and quarks. At the unification energy these otherwise distinct sorts of particle should merge in identity. At the relatively low energies at which we conduct particle physics experiments, the identity mixing is minute, but possibly detectable. The most dramatic consequence of lepton–quark mixing is the prediction that protons are unstable, and can decay. In one decay scheme the proton's down quark turns into a positron while one of the two up quarks transmutes into an up antiquark. The antiquark together with the remaining up quark constitute a pion.

The search for proton decay events is a key test of GUTs. Unfortunately the predicted lifetime of the proton varies from about 10^{28} years up to much longer durations depending on which GUT is used. Direct detection of proton decay with a lifetime much greater than about 10^{33} years is probably likely to prove technically intractable, so failure to detect proton decay would only be able to rule out some GUTs.

The technique used to search for decay events is to monitor a large mass of material for emitted particles. Proton decay, as with all quantum events, is a statistical process, so an average lifetime of, say, 10^{32} years implies a reasonable chance of detecting a single decay event in about a year in a mass containing 10^{32} protons.

Several such experiments have been performed. One of these, in a salt mine deep below Lake Erie (a location chosen to minimize the effects of cosmic radiation which would otherwise swamp the process of interest) employs a large tank of water in which are suspended an array of photomultipliers. A fast charged particle emitted during proton decay will give rise to a characteristic pulse of light (known as Cerenkov radiation) as it ploughs through the water. The aim of the experiment is to detect this secondary radiation. At the time of writing no unambiguous proton decay events have been recorded in any of these experiments.

Another potential test of GUTs comes from a quite different area: magnetic monopoles. All known magnets are dipoles; that is, they consist of north and south poles together. This is because in all cases the sources of magnetism can be traced to circulating currents, such as the motion of atomic electrons. A current flowing around a loop produces a north pole on one side of the loop and a south pole on the other side. Magnetic 'charge' as such has never been observed in nature. A magnetic charge would appear as an isolated north or south pole, more usually known as a magnetic monopole.

In spite of the lack of observational evidence for magnetic monopoles, Paul Dirac examined how they could be incorporated into quantum physics. In a classic paper of the early 1930s he found that, if magnetic monopoles exist, their magnetic charge m would have to bear a simple relation to the fundamental unit of electric charge e. Explicitly, $em = \hbar$, or an integral multiple thereof. This curious result implies, among other things, that if there is just one magnetic monopole in the universe, the value of e is fixed everywhere, which could explain why electric charge always comes in discrete multiples of a fundamental unit.

Dirac's work, however, gave no clue about the other properties that a hypothetical magnetic monopole would possess, such as its mass, and for many years physicists were inclined to suppose that the magnetic monopole was one of those particles permitted by the laws of nature but which nature chooses not to employ. This view altered dramatically with the inception of GUTs. Such theories not only have room for magnetic monopoles, they actually require them. Furthermore the theory supplies further important details about their likely properties.

The expected mass of a monopole is comparable with that of the X particles, i.e. about 10^{15} proton masses. This is so huge (about the same as a bacterium) that it explains why monopoles are not created in particle collision experiments. On the other hand, the necessary energy may have been present during the primeval phase of the universe, so some scientists have searched for relic cosmic monopoles left over from the big bang.

If cosmic monopoles exist, and bombard the Earth along with the other cosmic radiation, they would produce a number of distinctive effects. For example, a monopole striking an atomic nucleus could cause proton decay. Monopoles would also have a distinctive electromagnetic signature. If an electric current is made to flow around a ring of superconducting material, the magnetic flux threading the ring is found to be quantized, i.e. to have values that are discrete multiples of $\hbar$. If a monopole should pass through the ring the flux would jump by a fixed number of units by electromagnetic induction. It is sufficient for the experimenter to leave the ring in the superconducting state and hope for a monopole to chance along. In spite of a spectacular false alarm which occurred on St Valentine's day in 1982, no monopoles have been detected by these experiments or any other.

1.11 Supergravity

In spite of the encouraging progress made in the 1970s with schemes that unified the electromagnetic, weak and strong forces, gravity remained the odd man out. Gravity theorists were, however, by no means idle during this period. In the mid 1970s they made an important extension to the concept of supersymmetry. Recall (see Section 1.8) that supersymmetry is basically a geometrical symmetry, albeit of a rather abstract variety. Now Einstein's general theory of relativity is, of course, a geometrical theory of gravity. Several people discovered independently that supersymmetric geometry could also be used as the basis of a geometrical theory of gravity. The resulting theory became known as *supergravity*.

Supergravity incorporates Einstein's general theory of relativity but extends it. Einstein's theory remains true as an approximation, so that the excellent agreement between that theory and observation is not compromised. The main additional feature of supergravity is that the graviton is no longer the only messenger particle responsible for transmitting the gravitational force. Supersymmetry, remember, provides a link between fermions and bosons. If one applies a supersymmetry operation (a mathematical operation that involves rotating from ordinary dimensions to extra fermionic dimensions – see page 45) to a graviton, which has a spin of 2, the theory now describes a particle with spin $\frac{3}{2}$. No fundamental spin $\frac{3}{2}$ particles are known in nature, so this is something new. The new particles are called *gravitinos*, and there would be anything from one to eight different types depending on the particular form of the theory being used. Gravitinos, if they exist, share with gravitons the property of being exceedingly weakly interacting, so they would be very hard to detect experimentally.

Further supersymmetry operations produce many more particles, with spins 1, $\frac{1}{2}$ and 0. In the favoured supergravity theory, called '$N = 8$' on account of the fact that there are eight gravitinos, the total assemblage of superpartners to the graviton is 172. Attempts were made to identify some of these superpartners with the known particles of high energy physics, thus providing a possible superunification scheme. In this all-embracing conception, the messenger particles of the other forces – the photon, gluons, W and Z – would all belong, along with the graviton, to the same gigantic superfamily, a multiplet of particles linked together by means of supersymmetry. Thus might all the forces be unified, each force merely manifesting but one aspect of a single supersymmetric superforce. But this would not be all. Because the superfamily contains fermions too, these could perhaps be associated with the quarks and leptons – the fundamental particles of matter. Thus might matter and force be joined into a single theoretical description.

In spite of the compelling nature of such a grandiose structure, the identification of the graviton's superpartners with known particles remained a mere idealization. Nevertheless, many theorists were sufficiently excited to proclaim that supergravity might be the long sought after

Theory of Everything. Stephen Hawking of Cambridge University, in his inaugural lecture on taking up the Lucasian Chair of Mathematics, spoke of 'the end of theoretical physics being in sight' given the great promise of $N = 8$ supergravity.

Much effort was expended in refining the theory and exploring its ramifications. Supersymmetric versions of other field theories which are easier to analyse than gravity, were also developed for use as analogies. In one important development it was discovered that the geometrical structure of supergravity is considerably simplified if the theory is recast in more than four spacetime dimensions. The most favourable dimensionality for $N = 8$ supergravity is eleven.

While some theorists were busy reformulating supergravity as an eleven-dimensional theory in the early 1980s, in a parallel development others began studying extra dimensions in the context of the Kaluza–Klein theory. Their aim was to extend the original theory, which involved only gravity and electromagnetism, to incorporate the weak and strong forces too. This became possible because the theories of Weinberg and Salam and QCD had provided the weak and strong forces with gauge field descriptions closely analogous to electromagnetism.

In the original version of Kaluza and Klein's theory, electromagnetism was incorporated by appending just one extra dimension to spacetime, making five in all. This is related to the fact that only one sort of photon is needed to convey the electromagnetic force, which is in turn related to the fact that the gauge symmetry of the electromagnetic field is of the simplest variety (the $U(1)$ symmetry). The weak and strong forces, on the other hand, have more complicated gauge symmetries ($SU(2)$ and $SU(3)$) and require a multiplicity of messenger particles to convey them. This demands more than one extra dimension apiece in the extended Kaluza–Klein theory. When the whole package is put together, it turns out that spacetime must, once again, have a total of eleven dimensions.

In the eleven-dimensional Kaluza–Klein theory there is just one force – gravity. The electromagnetic, weak and strong forces are merely adjuncts of the gravitational force. The extended Kaluza–Klein theory is thus a completely geometrical theory of the forces of nature within a unified framework.

Here, the abstract gauge symmetries, so crucial to the successful formulation of a quantum field theory, are identified with the geometrical symmetries of the higher dimensions.

The coincidence that eleven dimensions emerge from both supergravity and Kaluza–Klein theory seemed very suggestive, and physicists began talking seriously about a Theory of Everything that employed both supersymmetry and higher dimensions. The extra dimensions, which were purely a mathematical device when originally applied to supergravity, came to be regarded as real physical dimensions, all rolled up to a minute size in the fashion prescribed in Kaluza and Klein's original theory.

Unfortunately, the eleven dimensional theory suffered from a flaw, which proved to be fatal. A distinctive element of the weak interaction is that it breaks left–right mirror symmetry (i.e. it is parity-violating – see page 39). This implies that elementary particles must be endowed with a definite handedness, or 'chirality'. In daily life we take the difference between left-handedness and right-handedness for granted, but the existence of chirality is actually dependent on deep properties of three-dimensional space. It turns out that definite chirality only exists in spaces with an *odd* number of dimensions. This means that space must have an odd number of dimensions, and hence spacetime must have an even number of dimensions, otherwise there would be no chirality in the laws of nature. In short, eleven spacetime dimensions won't work.

1.12 Mathematical diseases

In the foregoing we have made several references to problems of mathematical consistency in the formulations of quantum descriptions of the forces. In this section we shall look in a little more detail at the nature of these mathematical problems.

The first hint of the difficulties with quantum field theory actually arose in classical electromagnetic theory. One difficulty concerned the structure of the electron. A primitive picture of an electron is a tiny solid sphere with electric charge distributed uniformly throughout it. Because like charges repel, the charge in one region of the electron will repel the charge in the other regions, and there will be an outward force trying to blow the electron apart. Because of the inverse square law, this force will be very intense if the radius of the electron is assumed to be very small.

To prevent the disintegration of the electron, some other internal forces must also act. These cohesive forces are required to exactly counterbalance the disruptive tendency of the electric charge at all times, irrespective of how the electron moves. Attempts to model this balancing act in a fashion consistent with the special theory of relativity proved hopeless. Physicists eventually decided that the problem had to be sidestepped by assuming that the electron is, in fact, point-like; that is, it has zero radius and hence no internal parts to which a mechanical theory is applicable.

This notion served to solve one problem only to introduce another, for now there was a difficulty over the electrostatic energy of the electron. The energy required to assemble a charge on a sphere of radius r is proportional to $1/r$, so that if r is allowed to become zero, the energy is infinite. In the special theory of relativity energy has mass, which implies that the electron ought to have an infinite mass on account of its infinite electrostatic self-energy.

Although the presence of an infinite, or 'divergent', term in the equations is a severe embarassment, it need not be disastrous if the term concerned does not itself refer to a measurable quantity. In non-gravitational physics energy as such is not measurable, only energy differences. One is therefore free to rescale the zero of energy by an infinite amount so that the *observed* mass of the electron is finite. This rescaling is known as *renormalization*, and a theory in which finite answers are obtained in spite of the presence of infinities at intermediate steps is called *renormalizable*.

In the 1930s work began on quantum electrodynamics, a theory of electrons in interaction with photons, the carriers of the electromagnetic force. In QED the issue of the electron's electromagnetic self-action is more subtle. Indeed, the difficulty with infinities turns out to be more severe than in the classical theory. In QED electromagnetic forces are transmitted by the exchange of photons. Self-action arises in this description as a result of a charged particle emitting and then reabsorbing its own photon. Although this is hard to visualize, the Heisenberg uncertainty principle relieves us of the necessity to imagine the said photon actually turning around; the position and motion of the photon are smeared. The process is depicted schematically by the Feynman diagram shown in Figure 17.

The wiggly photon loop represents electromagnetic energy surrounding the electron. This energy contributes to the mass of the electron exactly as in classical electrodynamics. If the electron is again assumed to be point-like, there is no limit to the quantity of energy carried by these photons. The reason can be traced to the uncertainty principle for energy. The shorter the distance that the photon has to travel, the shorter is the time that it takes and the greater the energy uncertainty. For a point particle, the out-and-back journey need take no time at all, and the photon can possess infinite energy. Calculations show that the electron acquires an infinite mass from the photons that surround it.

But this time the renormalization trick is much harder to perform. First, other infinite quantities (such as electric charge) appear in the theory too, and these have to be taken care of. Second, Figure 17 represents only one infinite contribution to

Figure 17. An electron emits and reabsorbs a photon. Such processes 'dress' the electron with a cloud of electromagnetic energy. When the total energy is computed it turns out to be infinite.

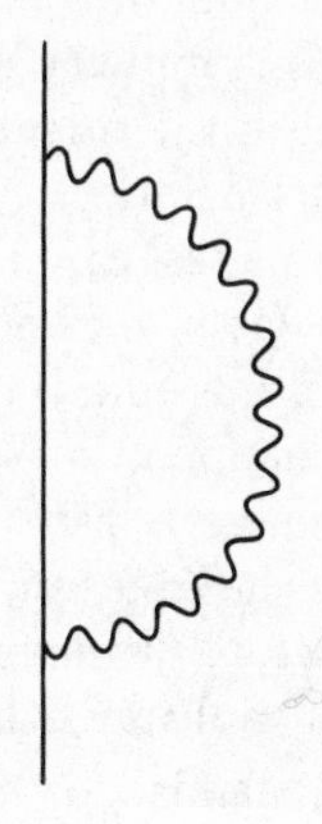

the electron's mass. Divergent terms also arise from the emission and reabsorption of two, three, four,... photons. In fact, there is an unending series of infinite terms. On the face of it, this seems to require an unending series of separate renormalization operations. If that were the case the theory would obviously be useless. It required a major mathematical investigation to demonstrate that, in fact, a single type of renormalization will remove *all* these infinities collectively from measurable quantities. Indeed, it took nearly two decades before QED could be declared renormalizable. It is a rare and significant property, and one which depends crucially on the gauge symmetry of the theory.

Quantum electrodynamics is not the only renormalizable quantum field theory known, but it is by far the most important. Its predictions have proved remarkably accurate, and it has been used as a prototype for other theories of forces. In contrast, the old theory of the weak interaction is not renormalizable, nor is the quantum theory of gravity based upon Einstein's general theory of relativity. In both cases, infinities recur endlessly, robbing the theory of predictive power and internal consistency.

Closely related to the problem of infinities is that of *anomalies*. An anomaly is the somewhat innocuous name given to the breakdown of some symmetry when a theory is 'quantized', i.e. when the classical theory is reformulated according to the rules of quantum mechanics. The presence of an anomaly means that a symmetry of the classical theory you start out with is violated in the quantum version of the theory. Because of the close association between symmetries and conservation laws, anomalies can result in the violation of sacrosanct conservation laws, e.g. energy and electric charge may not be conserved in the quantum theory. A crude way of understanding how this happens is as follows. If a quantity Q is conserved, its rate of change is zero. The quantization process often involves, as we have seen, some quantities being multiplied by infinite factors. It may then happen that the rate of change of Q becomes multiplied by infinity. We then have the product $0 \times \infty$. As it stands this expression is meaningless, but properly defined it can represent a finite quantity. This is what happens when an anomaly is present: the rate of change of Q becomes nonzero, and the conservation of Q is violated.

1.13 String theory

The search for a unified theory – perhaps even a TOE – had reached a curious and frustrating phase in the early 1980s. Certain ingredients, such as supersymmetry and higher dimensions, seemed to offer promising avenues of investigation. The considerable problems of infinities that had plagued all attempts to construct a quantum theory of gravitation seemed, if not to be absent, then at least to be ameliorated in the supergravity theory. And the extended Kaluza–Klein theory provided a compelling framework for amalgamating the four forces, even though it didn't quite work out properly. In short, theorists were very receptive to unification schemes that combined supersymmetry and higher dimensions. It was at this point that they began to take notice of string theory.

The origins of string theory go back to the late 1960s, and the work of Gabrielle Veneziano. At that time many physicists were trying to make sense of the proliferation of hadrons, the strongly interacting particles that had emerged from high energy collisions in particle accelerators. This was before the quark model of matter had been established.

Especially puzzling were the very short-lived hadrons with lifetimes of the order 10^{-23} seconds. These are known collectively as resonances, because it was quite obvious that they were not elementary particles; rather they seemed to be some sort of excited states of other hadrons. One could imagine the internal components of the hadrons being excited to higher quantum energy levels by the shock of high energy collisions. Investigation showed that some of these objects had very high spin (e.g. $\frac{11}{2}$). Moreover, a systematic relationship was found between the spin and the mass of these hadrons.

In order to explain these facts, Veneziano proposed an *ad hoc* model. At the time it was merely a mathematical procedure without any underlying physical picture. In the course of subsequent investigation, however, it became clear that Veneziano's model was describing the quantized motion of a *string*. This constituted a remarkable departure from previous theories, which had invariably modelled matter in terms of *particles*. Yet at least in some respects the string model was in better agreement with experiment than the more traditional particle approaches.

There is at least one sense in which a string theory of hadrons seems justified. As now understood, hadrons contain quarks, and these quarks interact through the inter-quark force. One can envisage the bonds produced by this force as being like pieces of elastic joining the quarks. Indeed, the inter-quark force shares with elastic the property that it grows with distance. As the quarks cavort about so the interactions within the hadron resembles a whirling string. In the case of quarks, the force is so strong that the interaction energy is comparable with the rest mass energy of the quarks. Under these circumstances the 'elastic' is more important to the dynamics than the quarks on the end of it. So modelling the motion just in terms of a string is not totally inappropriate.

The string model was never regarded as more than a crude approximation in those early days. Another problem was that it seemed to be restricted to describing bosons only. Nevertheless, a few people studied the model intensively and found some interesting results that hinted at the power of the theory. Then in 1970 John Schwarz and André Neveu discovered a second string theory that described fermions.

About 1974 QCD was developed, and the string theory ceased to be of interest as a model for hadrons. There it might have died, had not Schwarz and his then collaborator Joel Scherk spotted that it could be used in an altogether different, and much more exciting context. One of the problems of early string theory was that the particles it seemed to be describing included one with zero mass and a spin of 2. This didn't correspond to anything in the hadronic collection. It did, however, precisely describe the graviton – the messenger particle of gravity. Could it be, suggested Scherk and Schwarz, that string theory is really a theory of gravity? Could it even be a Theory of Everything?

It took a decade for this audacious idea to gain wider credibility. During that time a small group of theorists including John Schwarz and Michael Green dealt with all sorts of mathematical consistency problems – tachyons, infinities, anomalies, the need for higher dimensions, the need for supersymmetry. Ironically, they were widely regarded as wasting their time on an utterly crackpot theory. Today, all that has changed. String theory – in its modern guise known as *superstring* theory – commands the attention of some of the world's finest theoretical physicists.

In the coming chapters, some of the pioneers of string theory, including Schwarz and Green themselves, desribe the string model in detail. They give something of its history, its present status, and how they foresee future developments. They also address the crucial question of whether or not string theory really could amount to a Theory of Everything.

There is no doubt that string theory is extraordinarily compelling. Theorists speak eloquently of the incredible beauty and richness of the theory. But clearly another incentive for studying the subject derives from the fact that if superstring theory does eventually provide a quantitative explanation for all the particles and forces of nature, it will represent one of the greatest scientific triumphs in the history of mankind. Indeed, one might claim that it would be the culmination of reductionist science, because we would at last have identified the smallest entities from which the world is built and elucidated the fundamental principles on which the universe runs. No wonder that some workers abandoned longstanding research projects overnight to take up string theory. At the time of writing there is a veritable industry at work on the subject. In the fields of particle physics and gravitation theory scarcely a seminar is given or a journal published that does not refer in some way to strings.

Yet not all physicists are happy with this phenomenon. Some argue that the efforts of string theorists are philosophically and scientifically misguided. Some even say that the theory is plain baloney. Such criticisms receive their airing in the interviews that follow. Judge for yourself who you think will prove to be right. One thing is agreed: the stakes have never been higher for any scientific enterprise.

2

John Schwarz

John Schwarz is Professor in the Department of Physics at the California Institute of Technology. His early work, especially that in collaboration with Michael Green, advanced the subject from a theoretical backwater to today's powerful superstring theory.

The idea of using strings to model fundamental particles is one that goes back quite a way. Can you tell us a bit about the early days of string theory?

String theory has a very bizarre history. The subject originated in an attempt to solve a completely different problem from that which it's being used for today. Originally string theory was developed, around 1968–70, in at attempt to understand the strong nuclear force. It had a certain amount of success in that regard but was never completely successful, and in the mid-1970s another theory called quantum chromodynamics arose which was successful in describing the strong interactions. As a result, even though there was an enormous amount of work in string theory during that early period, most people dropped out of the subject in the mid-1970s when quantum chromodynamics was developed. The reason I didn't was because just prior to the development, or at approximately the same time as the development of quantum chromodynamics, I was working with a French physicist named Joel Scherk, who was visiting here at Caltech, and we noticed that one of the problems that we were encountering with string theory in trying to use it to describe the strong nuclear force, was that the theory always gave rise to a particular kind of particle that had no place in the strong nuclear regime. Namely this was a particle that had no mass, had two units of angular momentum, and just did not correspond to anything that was observed in nuclear processes. However, we knew that this was exactly the kind of particle that occurred in Einstein's theory of general relativity, which is a theory of gravitation, and this particle is usually called a graviton – the quantum mechanical particle which carries the gravitational force. Gravity is something very different from the strong nuclear force. Under ordinary circumstances it's much, much weaker, so since we were finding that this particle was occurring in the theory anyway, we decided we would abandon the project of using strings to describe the strong nuclear force and see if we could use it to describe gravity and at the same time other fundamental forces which, as it turns out, come along for the ride.

Turn a sin into a virtue in fact.

That's right. This required a rather dramatic change in viewpoint because it meant for one thing that the strings had to be much smaller than we originally conceived.

What sort of size are we talking about here?

When we were thinking of strings as describing nuclear particles, the idea was that the strings would have a size which was typical for a nucleus, which is 10^{-13} centimetres. When we use it for gravitation there is a natural length scale that is suggested by the structure of gravity. This is called Planck's length, and it is incredibly smaller than the nuclear scale – it's a factor 10^{20} smaller than that. One way that it is sometimes expressed is to say that the Planck scale is to the size of an atom as an atom is to the size of the solar system. We are talking about exceedingly small distances when we discuss strings used to unify gravity with other forces.

So the idea of using strings for gravitation and unification arose just in 1974 after string theory had already been developed for five years. Joel Scherk, who very sadly died six years later, and I continued to work on that problem, and in

1979 I began collaborating with Michael Green from Queen Mary College, London.

Before we go on to those developments, can I just ask you what sort of image did you have of neutrons and protons in the old string theory? Is it that in some sense a string is supposed to reside inside each neutron and proton?

Well, roughly speaking, the picture was that hadrons, such as neutrons and protons, were made up of quarks, a concept introduced some twenty years ago by Gell-Mann and Zweig. These quarks have to be held together by some force so the picture was that the strings were a description of the force that was holding the quarks together, like bits of elastic. One could think of the quarks as being attached to the ends of these strings.

And the whole lot whirling around in some way?

That's right.

What were the main difficulties with that idea?

There were several difficulties. One I have already mentioned, this massless spin 2 particle inevitably fell out of the mathematics and that is not part of the spectrum of particles that one finds in the nuclear regime.

Another difficulty, which is rather amusing, was that the mathematical consistency of the theory required that spacetime should have more than four dimensions. In the original string theory, which had other deficiencies as well, the theory led to twenty-six dimensions. In an improved string theory that was developed by Pierre Ramond, André Neveu and myself in 1971, that number was reduced to ten dimensions, and it is in fact a variant of that ten-dimensional theory that's in vogue today. Having extra dimensions was a very serious problem in the context of describing nuclear particles because we know darned well that there are three dimensions of space and one of time, and there is simply no room for extra dimensions if you want a realistic theory.

Did you hope that some reformulation of the theory might lead to a consistent theory in four dimensions?

Well, many efforts were made over the years – I devoted some fraction of my efforts to this as well – to try to find variants of these two string theories that would have four dimensions rather than ten or twenty-six. There were some interesting suggestions made along the way. They always started with a very mathematically beautiful system and made it really ugly and unconvincing and inevitably led to mathematical inconsistencies.

One of the other problems with the original string theory was the existence of so-called tachyons, particles that move faster than light. Was that inevitable?

In the twenty-six-dimensional bosonic string theory, that is an inevitable feature. One of the virtues of the ten-dimensional theory is that it's possible to choose a version of that theory which doesn't have any of these tachyonic particles, which we know are inconsistent with fundamental principles.

There were some successes of the old string theory as well, presumably.

Yes. This theory was developed for good reasons. It captured many general features that we knew we wanted in a theory of the strong nuclear force – certain features about how particles interact at high energies and other related features about the masses and angular momentum of the different particles and patterns that one can describe relating them.

Today, looking back at that phase, is it fairly true to say that one wouldn't turn to strings as a description of nuclear particles any more, that this has now been superceded by quantum chromodynamics?

Quantum chromodynamics is universally recognized in the community as being the correct theory of the strong nuclear force. I think the evidence for that is very compelling. However, it still seems quite plausible that one could reformulate quantum chromodynamics in a manner such that strings would be regarded as playing an important role. However, the strings that would arise in that context would have a different mathematical behaviour from the ones proposed fifteen years ago. The exact structure of such a string theory is only vaguely suggested by what we know today. In fact, it seems to be a much more difficult problem than the one that sounds so much more ambitious – the superstring theory that we are working on today.

What was the real turning point in the fortunes of the string theory, the thing that projected it into the forefront of research in particle physics?

It began with my collaboration with Michael Green in 1980, when we picked up on the work that I had started earlier with Jöel Scherk, developing the detailed mathematical behaviour of the ten-dimensional string theory. One of the important features of that ten-dimensional string theory that I should refer to is that it has a very special kind of symmetry called supersymmetry, which relates two different classes of elementary particles called bosons and fermions.

Could you say a little bit about what these two sorts of particles are?

All elementary particles fall into two different categories and these two types of particles, bosons and fermions, are different in two important respects. The amount of angular momentum that they carry is commonly described as 'spin', and the spin of a boson is an even multiple of a fundamental unit whereas for a fermion it's an odd multiple of that same fundamental unit.

Another distinction, which is very much tied up with issues of quantum theory, has to do with the behaviour of the theory when identical particles are interchanged, whether the theory is left invariant under this interchange, or whether it acquires a minus sign. The fermions give rise to this minus sign.

You are saying that supersymmetry is a means of amalgamating these two types of particles into a common description.

Yes that's right. Maybe to make this less abstract I should just mention that quarks and electrons are examples of fermions while photons and gravitons are examples of bosons.

Would it be true to say that we can think of fermions as being particles of matter, and bosons as being particles that convey the forces between the particles of matter?

I think that's a good way of putting it.

So you were saying that supersymmetry is an essential element in the modern version of string theory. What developments did that lead to?

Well, it's a long chain of ramifications. In fact, the discovery of the ten-dimensional string theory back in 1971 was really the birth of supersymmetry theory. One aspect of that was the supersymmetric generalization of the theory of gravity, which is a theory called supergravity that was worked out in 1976, and which is incorporated in the supersymmetrical string theory, more colloquially called superstring theory.

In studying the properties of the supersymmetric string theory, Michael Green and I found a number of things over the years that we thought were quite exciting. One of the significant problems that one has traditionally had in making a theory of gravity, is that when one tries to reconcile it with the requirements of quantum theory, the calculations always give rise to meaningless, divergent expressions. It's sort of analogous to dividing one by zero, which is an operation you just can't carry out. So one was getting these ill-defined answers in trying to do quantum calculations in gravity. This seemed to be a generic feature of all theories in which the fundamental particles were regarded as mathematical points, which is the traditional way of doing things.

So the important feature of string theory is to replace these points by one-dimensional curves called strings, and one exciting thing that we found was that when we calculated the quantum corrections to gravity for string theory, we were beginning to get numbers that did make sense, numbers that were given by finite expressions. This seemed to be the first indication that it was possible to make a finite theory that was consistent with quantum mechanics and contained gravity. That was exciting. We did that work around 1982.

At about the same time we found a couple of other superstring theories. One superstring theory that we developed contained what we call open strings (they have free ends). The other sort of strings form loops and are called closed strings. So the original superstring theory had both open and closed strings, but we discovered after a while that it was possible to make theories that had only closed strings. That turned out to be an important distinction because today it's the theories with only closed strings that look the most promising. In fact, they are actually simpler to study in many respects.

One of the important facts about nature that we want to account for in our fundamental theory, is that there is a distinction between left- and right-handedness. The theory should not be mirror symmetric – this is referred to as parity violation. It's an important property present in the standard model of the weak and strong interactions that we know applies at low energies. It's a challenge to understand this asymmetry from a more fundamental point of view, in particular in the context of superstring theory.

It turns out that of the few superstring theories which have been developed, all but one of them has this left–right asymmetry as a fundamental feature already in ten dimensions. So that's very encouraging. However, theories that have this left–right asymmetry have an enormous tendency to break down and give inconsistent answers, not the infinities I was referring to earlier, but a closely associated problem called anomalies. Basically the idea here is that before one takes account of quantum mechanics, the theory possesses a certain fundamental symmetry property, and the question one has to address is whether the quantum mechanical corrections respect that symmetry or whether they break it. If the quantum mechanical corrections break the symmetry, then the theory is inconsistent, and it doesn't make sense. This inconsistency happens generically whenever you have a theory with left–right asymmetry. So while it was exciting to have theories of left–right asymmetry, it was also very threatening because they were likely to have these anomalies which would make them inconsistent.

In 1984, Michael Green and I did a calculation for one of these superstring theories to see whether, in fact, this anomaly occurred or not. What we discovered was quite surprising to us. We found that, in general, there was indeed an anomaly that rendered the theory unsatisfactory. Now there was the freedom to choose the particular symmetry structure that one used in defining a theory in the first place. In fact, there were an infinite number of possibilities for these symmetry structures. However, for just one of them the anomaly magically cancelled out of the formulae whereas for all of the others it didn't. So amid this infinity of possibilities, just one unique one was picked out as being potentially consistent.

Let me just clarify this. We have two types of diseases in the traditional field theory approach to fundamental particles and forces. One is the presence of infinite terms, and the other is the presence of these so-called anomalies which introduce unwelcome symmetry breaking when the theory is quantized. Both of these make the theory mathematically inconsistent, but with the superstrings both of these problems seem to be taken care of provided one works with this uniquely selected version of the superstring theory. In what way is the theory uniquely selected? What feature is selected by this?

I mentioned that the particular symmetry structure was selected from an infinite number of possibilities that we had before the anomaly question was investigated. The name of the symmetry structure is called $SO(32)$.

We also discovered at about the same time that there was a second symmetry structure that seemed to be a consistent possibility which has the name $E_8 \times E_8$. The bizarre thing was that at the time, we didn't have a specific superstring theory that could contain that symmetry. So we had one superstring theory with one of the symmetries we had identified, and then we found a second symmetry that seemed like it could be consistent but we didn't have a theory to go with it. But shortly thereafter a group of four physicists at Princeton University, now known as the Princeton String Quartet, discovered two new superstring theories which they called heterotic strings. One of these theories incorporated the $E_8 \times E_8$ symmetry. The other one was a second example of a superstring theory based on $SO(32)$.

The $E_8 \times E_8$ theory was the one that aroused the most interest because that's the symmetry structure that looks most promising for accommodating the observed phenomenology of particles.

But already we seem to have quite a proliferation of alternative string theories. Isn't this bad news?

The numbers are still rather small. Ideally, of course, you're right. It would be best if there was only one possible theory and it explained everything. I think one could say that we have gone a long way in that direction, even if that isn't exactly the story as it stands today. At the moment there are, in ten dimensions, three heterotic string theories (the third one besides the two I have already mentioned was found later), and three superstring theories that are not heterotic – a total of six so far. However, it's possible that some of these theories will, on further investigation, actually turn out to be inconsistent. That would reduce the number. Also, it seems likely that the three heterotic string theories are actually different verions of the same theory. They can be shown, in fact, to be equivalent, so they should really be regarded as one theory. So by reasoning of that type there is a good chance that this number can be narrowed down to just a unique theory.

Why is it no longer a problem that these theories must be formulated in more than four spacetime dimensions?

As soon as we abandoned the hadron programme, the programme of describing the strong nuclear force with strings, and embarked on the problem of describing gravity and other forces, extra dimensions became an advantage rather than a disadvantage. The reason is that theories of gravity describe the geometry of space and time. So in the context of a gravitational theory, it makes perfect sense to suppose that extra dimensions do exist but that they are curled up into some tight little ball as a consequence of the geometry that is dictated by the theory itself.

So the theory will have extra dimensions. But the theory will also tell you what to do about them, because when you try to solve the equations, if all goes according to plan, you should discover that the solution of the equations entails curling up these extra six dimensions into a little ball sufficiently small that it wouldn't be observed.

How small are we talking about here?

The same length scale that I mentioned earlier – the Planck length – seems to be indicated. That's this incredibly small distance, 10^{-33} centimetres.

You're saying that each point in space, or what we thought was a point in space, is, in fact, a little six-dimensional ball about 10^{-33} centimetres across. So it's not surpising that we don't notice these extra dimensions.

They are too small to be detected.

How should we then envisage these strings? Should we think of particles, such as electrons or quarks say, as in some sense made up of strings? Do we think of there being a little string inside them? A loop, something of that sort?

Well, I would express it a little bit differently than that. When you have a string it can oscillate and vibrate in different ways – rotate and so forth – and each of these different modes of vibration or oscillation can be thought of as describing a particular type of particle. So one can think of the electron as one mode of vibration, and a quark as another mode of vibration, and a graviton as yet another.

Just one type of string inside, but moving in a different way, a different pattern of motion?

Yes.

You mentioned that the most promising approach to the superstring theory was with the so-called $E_8 \times E_8$ formulation. What is the significance of these two versions of E_8?

It's not entirely clear how the story is going to look by the time the dust settles, but one possibility which seems intriguing at this time, is that the symmetries of particle physics, which we know about from experiments at energies that are currently accessible, are part of the symmetries of one of these two E_8s. The other E_8 symmetry describes a new kind of matter, sometimes called shadow matter, that doesn't interact, or only interacts extremely weakly, with the ordinary matter that we are familiar with. If you wanted to construct some science fiction out of this, you could imagine all sorts of galaxies and planets made out of shadow matter that would be completely invisible to us because they wouldn't interact with our kind of light.

So, one amusing possibility is that the shadow matter associated with this second E_8 symmetry would be essentially invisible to us because it wouldn't interact with the kind of light that we are able to detect.

Could there be shadow matter passing through this room at this very moment, and we wouldn't know about it?

That's true. There are some limits that you can set on it because it does interact with our kind of gravity – we share our gravity with shadow matter.

So we would notice a shadow planet?

We would notice it through its gravitational effects although we wouldn't see it with light.

Is there any evidence at all that shadow matter exists?

No, there isn't. It's consistent, however, with what we know about the universe because there is evidence that the visible matter in the universe only accounts for maybe ten percent or so of the total mass of the universe. So even if half the mass of the universe were in shadow matter, that would be quite consistent. There is room for it.

Would this shadow world be more or less identical to our world in the nature of the particles and their interactions?

That's a question that depends on the details of how the theory works out. One possibility is that the two E_8 symmetries each break down in an identical manner to smaller symmetry structures. If that symmetry-breaking pattern was the same for each E_8 factor then you would have the same symmetry in the rules of physics for the two kinds of matter. At the moment it seems more likely that this symmetry breaking takes place in different ways for the two E_8s.

Why should that be? What distinguishes one from the other?

In trying to solve the equations of the theory, the only solutions that people have succeeded in finding so far are ones in which you invoke different symmetry breaking patterns for the two E_8s.

So it's lop-sided between the world and the shadow world?

Yes, but it's not out of the question that other solutions will be found in which they are treated symmetrically.

As I understand it, one of the major outstanding problems of the superstring programme is the question of deciding which particular configuration the six extra dimensions will curl up in. Do you see this as being an insuperable obstacle, or is it something that in a few years will yield to mathematics?

Well, it's an enormous challenge, and it's perhaps one of the two most fundamental questions in the subject today. If we knew what that six-dimensional space looked like we would be in a great position for calculating all sorts of things that we want to know. This may sound surprising. After all, as I have already said, this space is completely invisible because it's too tiny to observe directly. As it turns out the details of its geometry and topology actually play a crucial role in determining the properties of observable particles at observable energies.

Can you give an example?

There's one topological property of this six-dimensional space called its Euler number, which can be thought of, roughly speaking, as measuring how many holes are in it. This Euler number, it turns out, is directly related to how many repetitions there are in the families of quarks and leptons. It is found that quarks and leptons occur in groupings called families. Three of these families have been observed experimentally, and it's one of the mysteries why there are three families of quarks and leptons. One of the exciting facts about string theory is that the number of families that arises is just given by half of the Euler number of this six-dimensional space.

So here we would have an example of how the topology of these unseen dimensions of space would directly affect something physical, like the numbers of different types of particles found in nature?

Yes.

One of the problems with superstring theory is that at the present time there seems to be no unique theory but a very large number of theories depending on how one chooses to curl up these extra dimensions. Roughly, how many different choices have we got?

Let me first express the same sentiment that you did in slightly different words. I would say that the theory is unique and the non-uniqueness is introduced in the solutions – one theory has many different solutions – and our big problem is to try to understand why one solution would, in some sense, be better than all the others and, even better, explain nature.

At our present level of understanding there is no way of choosing among these different solutions, other than saying that one seems to fit nature better than another, but no mathematical criteria for saying that one is better than another.

However, the theory is not yet completely understood. We are still looking for the best formulation. In particular, in the present formulations of superstring theory, we are only able to study the theory in various successive approximations, a scheme called perturbation theory. What we are groping for is a formulation of the theory which won't tie us to this particular kind of expansion in terms of successive approximations. If we had a formulation of the theory which gave us exact results rather than successions of approximations, it's possible that we would discover that some of these six-dimensional spaces which appear to solve the equations to any level of approximation that we can presently study, won't in fact solve the equations when we look at them from an exact point of view.

So if you could do exact calculations you might actually single out one unique solution.

That's right. The jargon way of expressing this is to say that there could be nonperturbative effects in the theory which would rule out all but one, or some of the solutions.

But until you know about that, how many different contenders are there?

It's hard to really count, but I suppose you could say there are thousands, maybe more.

Apart from that, what do you see as the major outstanding problems of the theory?

Formulating this nonperturbative version of the theory, this exact description of the theory. We are in a peculiar position with string theory in that we know some of the equations but we don't really have a deep understanding of the principles that underlie those equations. The history here is backwards from the development of Einstein's gravity theory – the general theory of relativity. Einstein started with a very beautiful principle, called the equivalence principle, and from that principle he constructed some equations, which can then be studied.

In the case of string theory we have a certain set of equations, but we really don't understand the generalization of the equivalence principle that's responsible for those equations. But it's clear that there is a very deep and beautiful mathematical structure that underlies all of the startling results that we are finding, and that some very elegant and profound principle is there to be found. There has been a lot of work in the last year or two, trying to clarify what's going on, and some extremely recent work might actually be pointing the right way, but it's still very preliminary and needs to be studied more before we know for sure.

So there is not only the problem of trying to solve the theory and fit it to experiment, but also, at a more fundamental level, we have to deepen our understanding about what the theory is all about.

Assuming that the theory continues to be successful, where would you expect contact with experiment to be made? So far we have a very elegant formulation of things which are already known but to be properly successful a theory has to make some new predictions that can be tested.

That's certainly true, and it's impossible to know on what kind of a time scale such successes might be achieved, if they ever are. I'm hopeful that some compelling evidence for string theory will be found before the end of the century, but I certainly wouldn't make any promises to that effect. One really doesn't know how long it's going to take. We are asking very ambitious types of questions here, it's a very ambitous programme, and there's no guarantee that it's going to succeed at all, even though it does seem much more promising than any approach that has preceded it.

Is it likely that the theory will predict new particles that will be discovered by new accelerators?

Let's suppose that we succeed in understanding the fundamental principle, that we can find a unique solution to the equations. Given that solution we can study the topological properties of that six-dimensional space. From that we can learn the types of particles that would exist at low energy, the ratios of their masses can be worked out just from topological considerations, as well as the strengths with which they interact with each other.

This is the kind of information that one deduces from experiments that are carried out in the lab. Certainly there are particles that are yet to be discovered, associated with supersymmetry for example, or associated with symmetry breaking. At the present time we have only rough ideas as to what the masses of these particles and some other properties should be. If we had a specific compactification of the six dimensions that successfully accounted for what we already know, it would presumably at the same time make predictions for some of these other questions which could then be tested experimentally.

I get the impression that advances will only come if there is some major new development in understanding the mathematics behind this – that you are appealing to certain branches of mathematics which are themselves new and for which one needs further progress.

That's right. This is one aspect of the whole subject that a lot of people find kind of frightening, that these studies require an enormous amount of mathematics. In fact, much of this mathematics hasn't even been worked out yet by the mathematicians. There is a lot to learn, a lot of new results in mathematics to be developed, at the same time that we are trying to understand the physics. It's an exciting time to be involved in all this, and I'm optimistic that in the long run it's all going to pay off.

People refer to the superstring programme as a Theory of Everything because the ultimate goal of the theory is to explain all particles and all forces. It's often said that there have been periods in history when Theories of Everything have been just around the corner. So far that's always been wrong. What right have we to suppose that the superstring theory is going to be different?

Well, in all previous (partially) unified theories that have been successful, one has described some of the particles and forces that were known at the time and not others. This is a programme that hopes to account for all of the forces, including gravity. There have been enormous successes with the unification of the electromagnetic and weak forces in recent years, and then extending that to include the strong force. That work is very exciting and successful, but could never claim to be all-emcompassing because it is clear that it is leaving gravity out of the picture.

Other approaches that have been proposed in the past for describing gravity didn't have any chance of accounting for the other forces. So this is the first instance (that I'm aware of at least) of a programme which contains gravity and looks like a plausible candidate for describing the other forces at the same time. It's such a tightly-knit mathematical structure that it's not something you can really vary at all. If it succeeds in describing experimental results, it's hard to imagine that this kind of a theory would be an approximation to some yet better theory to be discovered in the future. It's such a tight structure that if you do anything to it I think it would completely fall apart.

So in that respect, it really differs from theories that have preceded it. In the past theories have always been regarded as some low energy approximation to something deeper that lay in the future.

Looking on the optimistic side, just supposing it all works out very nicely and it's possible, maybe by the end of the century to make detailed predictions about things that we can observe, and people become very confident that superstrings are describing the fundamental principle on which the world is built. What then of theoretical physics? Will it have come to an end?

I think that's a logical possibility, but very unlikely. The field of elementary particle physics is different from all other branches of physics and science, in my opinion, in that it asks a very specific question, namely what are the fundamental particles and forces of nature and what are the rules that govern them? This is a question that in principle one could find the right answer to, and then be done. All other fields of science, it seems to me, are open-ended. You can always ask new questions.

In that sense, then, what you hint at is something which is logically conceivable. However, our experience to date is that whenever we succeed in answering a question, five more get raised by the knowledge of that answer. We see no indication that that's not going to continue to be the case for a long, long time.

So while we are hopeful of having a complete understanding of fundamental particles and forces – and that may eventually be achieved – I think it's going to take much more than fifteen years, say, to do that even though in such a time span one could imagine that enough successes might be achieved to make it convincing that one is on the right track.

An assumption that underlies all approaches to describing the fundamental particles and forces of nature is that we can, in a simple way, image nature through bits and pieces of mathematics, and we hope ultimately simple mathematics, or at least elegant mathematics. Is this just some sort of pious hope or do you think the world really is built out of simple mathematical principles?

That seems to be the case, and it's rather a deep philosophical issue why this should be the case. I don't have the answer to it. It seems reasonable that there is some logical explanation for everything and mathematics seems a way to describe things in a logical manner. This faith, it seems to me, is based mostly on our experience that mathematics has had enormous successes in describing nature so far, and all the time this continues to be the case on a deeper and deeper level. So it will just be extrapolation, I suppose, that this is going to continue.

Of course, it could be the case that when you reach a certain level, say the level of the subatomic particles we are now investigating, things look temporarily very simple but if one digs deeper it gets very complicated again.

Well that's a sentiment that many people have expressed. If superstring theory doesn't work out then that might be an alternative viewpoint to adopt.

Another concern one might have is that the mathematics which is required just gets so difficult that the human mind is unable to deal with it! That's a fear we have from time to time.

I have heard it said that the superstring theory is the last hope of getting a Theory of Everything, at least a theory based upon simple or tractable mathematics. Do you think this is so?

I don't know if that's so or not. I think people have thought that way about other theories in the past, and if for some reason it doesn't work, I would suppose that some other candidate would then be propounded.

To finish on a personal note, when did you first realize that you were on to something big?

During my collaboration with Michael Green, which as I already mentioned started in 1980, we made several discoveries, one or two each year that we felt were of some importance, and with great enthusiasm we would publish them and talk about them in lectures to our colleagues around the world. In each case I felt, and I suspect Michael did as well, that this was the discovery that was going to convince people that this was really an important subject to be pursuing. So it was rather surprising to me that for quite some years the rest of the theoretical physics world was really not particularly interested in this work, or at least if they were they didn't show it! They were polite to us, they tolerated us, but certainly other people didn't start working on it.

When we found the anomaly cancellation in the summer of 1984, I had already become sufficiently accustomed to the way the community was responding, so I didn't expect anywhere near the kind of enthusiastic reaction that the work in fact got. I always expected that superstring theory would eventually become the important fashion for unification, but I expected the transition to be rather gradual. In fact, after the summer of '84, it was less than a year before a large number of people were working on this.

How does it feel now, immersed in this whirlwind of activity that has overtaken your subject? Do you feel that in some sense you can stand back and watch it develop? Obviously you are still very active in the subject.

I want to continue to be active and try to continue to make contributions. There are an awful lot of clever people doing very impressive work and it's not easy to compete with some of these people. Some of the younger guys especially know an awful lot of the modern mathematics that is required and are doing very well. This development, of course, pleases me enormously, because when only two of us were doing the work (although I should mention that we also worked with Lars Brink, so two and sometimes three), one of the frustrations was that the subject was developing rather slowly. There were a lot of interesting problems but we just didn't have enough time and energy and ability, I guess, to pursue them all, and we were anxious to see where it was all leading. Now it's developing so rapidly that it's impossible to keep up with the literature – I get a pile of new papers every day – and one could easily spend one's time just reading them and not doing anything else at all!

3

Edward Witten

Edward Witten is Professor in the Institute for Advanced Study at Princeton. He made many important contributions to theoretical particle physics and quantum field theory, especially in the area of quantum chromodynamics and higher dimensional theories, before turning to superstrings. He is one of the most lucid and outspoken advocates of the subject.

What are the essential problems that the superstring theory claims to address?

In twentieth century physics there are two really fundamental pillars, one of them is general relativity, which is Einstein's theory of gravity, and the other is quantum mechanics, which is the theory of everything that goes on in the microscopic domain. In other words it's the theory of atoms, molecules and smaller objects called elementary particles. The basic problem in modern physics is that these two pillars are incompatible. If you try to combine gravity with quantum mechanics, you find that you get nonsense from a mathematical point of view. You write down formulae which ought to be quantum gravitational formulae and you get all kinds of infinities. It's pretty discomforting for a physicist to find infinities in the middle of his calculations.

Most people who haven't been trained in physics probably think of what physicists do as a question of incredibly complicated calculations, but that's not really the essence of it. The essence of it is that physics is about concepts, wanting to understand the concepts, the principles by which the world works. In really fine theories like general relativity there is a well-defined, conceptual formulation and when you learn it you say 'yes those concepts are perfect', and the theory that is based on them is the best embodiment of those concepts.

Now quantum mechanics is a little bit different. It developed through some rather messy, complicated process stimulated by experiment. While it's a very rich and wonderful theory, it doesn't quite have the conceptual foundation of general relativity.

Our problem in physics is that everything is based on these two different theories and when we put them together we get nonsense. The history of physics is the history of discovering more refined concepts on which natural laws have been based. As these concepts get more refined a theory with fewer and fewer principles has to do more and more things at once, and it gradually becomes more and more complicated to write down something which is internally consistent. In Newton's day the problem was to write something which was correct – he never had the problem of writing nonsense, but by the twentieth century we have a rich conceptual framework with relativity and quantum mechanics and so on. In this framework it's difficult to do things which are even internally coherent, much less correct.

Actually, that's fortunate in the sense that it's one of the main tools we have in trying to make progress in physics. Physics has progressed to a domain where experiment is a little difficult and certainly it isn't developing as rapidly as it used to fifty or sixty years ago. Nevertheless, the fact that we have a rich logical structure which constrains us a lot in terms of what is consistent, is one of the main reasons we are still able to make advances.

So, the most important thing to bear in mind about string theory is that it aims to overcome what has been the central problem in physics for decades: the inconsistency between gravity theory and quantum mechanics.

How does it overcome this inconsistency?

A major headache for physicists throughout the twentieth century has been that if you take a particle like the electron and think of it as a point object, and then you take seriously its electric and gravitational fields, you find that there is an infinite energy in the electric field and an infinite energy in the gravitational field. Now this problem went through many different stages. It was a headache for classical physicists. In the case of the electric field it was a headache for quantum mechanicians, after quantum mechanics was developed.

The crucial step in the case of electromagnetism was that the uncertainty principle made the electron a little bit fuzzy and that enabled us to make sense of its electric field.

When it comes to making sense of the gravitational field of the electron we find that if we consider the electron to be a point particle, which is what most physicists have believed this century, it just doesn't work. However, in string theory the electron is no longer a point particle, but is a little vibrating string. That extra dimension of the vibrating string enables us to make sense of its gravitational field. I give the electron only as an illustration. The infinite electric energy of the electron is the classic problem of its kind, but we have a similar problem for all the elementary particles, and string theory neatly deals with this for all of the particles and all of the interactions.

So we no longer think of the world as made of particles at all, but as made of little strings that wriggle about?

That's right. When we think about particles we have to remember that since the inception of quantum mechanics everything in the world has been regarded as a little bit fuzzy, that is, a little bit fuzzy compared to the everyday image of what a particle is. In string theory this fuzzy quantum particle is replaced by a little quantum string. It's a vibrating string and on top of that it's made a little bit fuzzy by quantum mechanics.

Are there many different types of strings?

There are a few possible string theories but in most of the string theories there is basically one kind of string. You see, one kind of string can execute many different kinds of motion. If you think about a violin, a violin string when you play on it can vibrate at many different frequencies, called harmonics. The different harmonics of a violin string are essential to the richness of the sound, and that's the reason that different musical instruments sound different, even if you play the same note. You can play C on a piano or on a violin, and it will sound quite different because the same string can vibrate in different ways with different harmonics. The different instruments produce the different harmonics in different proportions.

In the case of a violin string, the different harmonics correspond to different sounds. In the case of a superstring, the different harmonics correspond to different elementary particles. The electron, the graviton, the photon, the neutrino and all the others, are different harmonics of a fundamental string just as the different overtones of a violin string are different harmonics of one string.

Is it stretching the analogy too far to say that the different fundamental particles of nature in some sense represent different musical notes?

That's a pretty good analogy.

How large are these strings?

The string corresponding to an electron, let's say, might be only 10^{-33} centimetres across so that's vastly smaller compared to things that you might well think of as unimaginably small. An atom might be 10^{-8} centimetres across, a nucleus is one hundred thousand times smaller but a superstring representing an elementary particle is unimaginably smaller even than that.

But nevertheless not a point object, that's the crucial thing.

No, it's not a point object, and the fact that it is not a point object but has a definite and positive size is essential to the consistency of the whole scheme. I might say, incidentally, that although a superstring representing an elementary particle is incredibly tiny, if you had sufficiently powerful tweezers there is no reason in principle why you couldn't grip one of these things and stretch it, making it larger and larger. Whether it would snap would depend on the particular string theory, but for most of them the string wouldn't snap, and you could actually make a wire stretching across the room that would be a macroscopic superstring. This would be analogous to another kind of string which is often discussed by physicists and astrophysicists nowadays which is the so-called cosmic string, some of which may stretch across the sky and perhaps be detected by astronomers.

Are you suggesting that there may be cosmic strings in the universe that are remnants of superstrings left over from the big bang?

There might be, but I don't want to emphasize it particularly. But in some of the string theories it would be quite possible in principle to have the strings stretching across the sky, that you would detect with a telescope.

Can you say something about the topology of the superstrings?

In most of the string theories they always form loops. All string theories include closed strings, strings that form loops, and most of the string theories *only* have closed strings, but one of the string theories, the so-called Type 1 string theory, has both open and closed strings.

What first attracted you to the string theory?

Mainly the possibility of reconciling gravity with quantum mechanics. It has been the central problem in physics for longer than I've been in the field. Quantum mechanics and quantum field theory were first developed in the late twenties. From the beginning it was clear that there was a problem of consistency between gravitation and quantum mechanics. In those early days quantum field theory had a lot of other problems and physicists didn't focus on this one so much, but in time, as the other problems were resolved, the inconsistency of gravity and quantum mechanics emerged more and more as *the* central problem in theoretical physics. Perhaps the most inaccessibly difficult problem. There were periods when it did not motivate a lot of work because it was too difficult and didn't give rise to a lot of interesting ideas.

String theory is extremely attractive because gravity is forced upon us. All known consistent string theories include gravity, so while gravity is impossible in quantum field theory as we have known it, it's obligatory in string theory.

That's only one side of what's so fascinating about string theory. Another is the remarkably rich mathematical structures that it gives birth to. I think that's very significant because over the years progress in physics has involved successively richer mathematical structures. I myself think it is no accident at all that progress in reconciling gravity with quantum mechanics has brought onto the agenda of theoretical physicists such a rich mathematical structure.

What areas of mathematics are being opened up by this theory?

Well, the theory of Riemann surfaces, the theory of certain kinds of symmetries called Lie algebras, and a variety of others. Many areas of mathematics that haven't been significant in physics in the past are quite important in string theory. This continues a process which has always occurred before when significant advances in fundamental physics took place.

The areas you've just referred to are branches of geometry, or it's generalizations. Is that correct?

These are mainly areas of geometry, and perhaps algebra. String theory at its finest is, or should be, a new branch of geometry. Einstein's great achievement in general relativity was to base the theory of gravity on geometry, Riemannian geometry to be precise. If string theory is to be a worthy successor of general relativity it must likewise have a geometrical foundation of which at present we only have glimmerings. But many of us are firmly convinced that it does exist.

Do you believe that many of the physical properties of subatomic particles, in fact, have a geometrical origin?

General relativity, in which one physical phenomenon, gravitation, is based on geometrical principles, is really in its way the most perfect and satisfying of physical theories. It's been the ambition of physicists since Einstein to achieve the same kind of perfection in other branches of physics, ultimately in the form of some unified picture of physics.

I, myself, believe rather strongly that the proper setting for string theory will prove to be a suitable elaboration of the geometrical ideas upon which Einstein based general relativity. Incidentally, I would consider trying to elucidate this proper generalization of geometry as the central problem in physics, certainly the central problem in string theory.

Do you think we will be able to understand qualities such as electric charge in geometrical terms?

I think that string theory in total will prove to be a geometrical theory and inasmuch as it succeeds in accounting for the various forces, it will give what could well be called a geometrical basis for properties such as electric charge.

You have been working on this theory now for a while and must have built up a sense of where the theory is going. How hopeful are you that it really is going to prove to be the fundamental Theory of Everything, to use a phrase?

I don't like to speculate about Theories of Everything, but what I will say is that I really believe that string theory is leading us to a fundamental new level of physics, comparable in scope to any of the advances that have been made in physics in the past. At the same time I think one has to regard it as a long-term process. One has to remember that string theory, if you choose to date it from the Veneziano model, is already eighteen years old, and looking back into the past we can see that ten or fifteen years ago there was a long road ahead, a lot of things that weren't known that had to be known, and it's probably still true today. We are probably living in what might prove to be the early stages of a long process like the process that led to quantum electrodynamics. Quantum theory started with Planck's work on black body radiation in 1900, and that original work of Planck really involved a formula in what we would call the quantum theory of electricity. Yet that quantum electrodynamic theory towards which Planck was heading, took fifty years to emerge.

So it's very probable that the eighteen year journey we've travelled so far is analogous to the early stages of the long process that led to quantum electrodynamics.

Are the implications going to be as profound as those that emerged from quantum electrodynamics?

I would expect that a proper elucidation of what string theory really is all about would involve a revolution in our concepts of the basic laws of physics – similar in scope to any that occurred in the past.

Given that the theory is still in its formative stages, can you point to any definite successes so far? Is it just pretty mathematics that looks exciting to physicists, or is there something more concrete?

Reconciling gravity with quantum mechanics is a remarkable achievement. That has been *the* problem of problems in physics since long before I was in the field.

Would you say this reconciliation is manifest at this time?

I would say so, yes. I would say that at this stage, and in fact for some years, it's been clear that string theory does, in fact, give a logically consistent framework, encompassing both gravity and quantum mechanics. At the same time, the conceptual framework in which this should be properly understood, analogous to the principle of equivalence that Einstein found in his theory of gravity, hasn't yet emerged.

I might remark that history shows us that reconciling inconsistent physical theories is a very good way of making fundamental progress. If we look at some of the advances in the twentieth century we see that Einstein's theory of special relativity came from a wish to reconcile two outstanding theories of the day, namely Maxwell's theory of electricity and Newtonian mechanics. Einstein's theory of general relativity likewise came from an effort to reconcile his own special relativity with Newtonian gravity. Finally, quantum field theory came from an effort to reconcile nonrelativistic quantum mechanics with special relativity. So many of the most far reaching advances in the twentieth century have come about because previous theories weren't compatible with one another. History teaches us that reconciling incompatibilities between theories is a good way to make really fundamental progress.

What do you think at this time are the main outstanding problems with the theory?

The purpose of being a physicist isn't just to learn how to calculate things, it's to understand the principles by which the world works. As I have indicated previously, physics is essentially a matter of discovering concepts. The outstanding thing which is dissatisfying in string theory at present is that despite its many remarkable features and a lot of wonderful discoveries that have been made, we have very little grasp of the proper conceptual framework for it, analogous to geometry in the case of general relativity. The central question on which we would most like to make progress would be to try to elucidate the logical framework in which string theory should be understood. The problem may well remain unsolved for many years.

General relativity springs in an inevitable way from the principles on which it is based. Once you grant that you want to base gravity theory on geometry, and you understand special relativity, from the few general principles which can be described graphically in physical terms (e.g. Einstein's famous elevator thought experiment and a few others), once you grasp the concepts, the mathematics follows. The mathematics is a perfect embodiment of those concepts. You could hardly improve upon it.

What we suspect exists in string theory, and it's what we would like to discover more than anything else, is an analogous, conceptual, logical framework in which string theory is as natural as general relativity is in its terms. We want to discover this because discovering the concepts by which the world works is the purpose of being a physicist anyway. We would also like to discover the right conceptual framework because it's very likely that a proper understanding of what string theory is all about is essential in order to do the calculations that we would like to do. We would like to use string theory to calculate the masses of elementary particles, coupling constants, lifetimes, interactions, and probabilities for all kinds of processes. It is only by doing such calculations and comparing with experiment that we can ever know if a theory is right.

But it is very likely that when a theory is only understood in a rather crude way, and the proper foundations for it aren't known, it might prove difficult to carry out these calculations. I think certainly the intellectual pay-off and quite possibly the practical pay-off require understanding of the logical framework. Certainly that's the problem in physics that I would wish to make some progress with if I could have my wishes.

Given that it's such a tough thing actually to calculate these masses, coupling constants and the low energy consequences of the theory, is it likely that there is going to be another area in which experimental tests can be carried out? Is the theory likely to predict the existence of new types of particles or new phenomena which perhaps will be revealed in particle accelerators?

I interpret the question as being if you could calculate everything in string theory then you could determine experimentally if it was correct in short order. Likewise if you could do experiments at the so-called Planck energy, where the basic problems of gravity and quantum mechanics become manifest, you would be able to quickly determine whether string theory was correct.

But that's hopeless, isn't it?

Unfortunately, it's hopeless to do experiments at such high energies and it is also hopeless, at present, to calculate everything. So there is the question of whether we can find some lucky corner of things where we can make an unusual prediction without being able to understand string theory comprehensively. It's conceivable, but I'm not too optimistic that it will happen within the next few years.

Are there no predictions of new particles, or anything of that sort?

Well, a lot of string theories, and a lot of models for how string theories work, will predict fractionally charged unconfined particles with almost Planckian masses which you could conceivably discover in cosmic rays.

We are talking here about particles which are immensely heavy compared to known fundamental particles?

We are talking about an individual elementary particle with a mass similar to that of a bacterium.

But with the extraordinary signature that its charge will be a fraction of that found on other particles?

Right. A fraction of that of the electron.

And these will presumably be left over from the big bang?

Our only hope of detecting them would be that they should be left over from the big bang. You can make a little estimate – how many of them could there conceivably be in cosmic rays? We know how much mass there is unaccounted for in the vicinity of the solar system, this is sometimes called dark matter. In the most optimistic case, if that unaccounted-for mass were made of these Planckian, fractionally charged particles, then the particles could be discovered in magnetic monopole detectors. That's an example of what I would call being lucky because no-one is going to tell you that the dark matter is actually made of such things. I'm sure there are a lot of other ways of being lucky that nobody has thought of yet, but I wouldn't want to hazard a guess on when we might enjoy such a stroke of luck.

You have been using the term 'string theories' in the plural and this seems to be inconsistent with claims that these theories are constrained. It's often said that one of the beauties of the string theory approach is that it doesn't provide you with too much room for freedom. Just how many string theories are there?

To put it in perspective, you should bear in mind that in conventional quantum field theory there are infinitely many possible theories. There are literally thousands of them that have been seriously contemplated by theoretical physicists. Compared to that, string theory is pretty good at present. There are about four or five, maybe six consistent string theories depending on how you count them.

What sort of criteria can be used to narrow this down?

For the time being we should get considerable satisfaction from having narrowed down millions, or perhaps thousands, or an infinite number of theories, to about five or six. If we didn't go further then we could still have immense satisfaction.

One of the peculiar, some people may think bizarre, features of the superstring theory is that the strings live not in the usual three dimensions of space and one of time that we perceive, but in a higher-dimensional universe. Is it reasonable for us to accept the existence of these extra dimensions?

Everything in the natural world is a little bit fuzzy because of Heisenberg's uncertainty principle and the basic ideas of quantum mechanics. If you've got some extra dimensions but they are so tiny that the fuzziness of everyday life blurs everything out on a size which is bigger than the size of the extra dimensions, then you would only notice the extra dimensions with extraordinary efforts. The idea is that if the extra dimensions are so tiny, then you just don't notice them.

I might say that the idea of extra dimensions might sound a little bit strange to anyone who hasn't studied physics. Anyone who has gone into physics professionally, will know that there are many things that are a lot stranger than extra dimensions. General relativity is strange, quantum mechanics is strange, antimatter is strange. All these things are strange but true. Compared to a lot of things that have come true in physics in the past, extra dimensions are not such a radical departure.

Can we understand how these extra dimensions have curled themselves up to such a small size?

We can *try* to understand it and we can see that by making some simple assumptions about how the extra dimensions would curl up, we can get plausible and interesting rough models of particle physics. I don't think we can expect to understand definitively how the extra dimensions curl themselves up without understanding a little better what string theory is really all about. We are handicapped by having an extremely primitive and crude view of what the subject really is.

Einstein developed general relativity at a time when the basic ideas in geometry that he needed had already been developed in the nineteenth century. It's been said that string theory is part of the physics of the twenty-first century that fell by chance into the twentieth century. That's a remark that was made by a leading physicist about fifteen years ago. What he meant was that human beings on planet Earth never had the conceptual framework that would lead them to invent string theory on purpose. String theory was invented essentially by accident in a long sequence of events, starting with the Veneziano model that was formulated in 1968. No-one invented it on purpose, it was invented in a lucky accident. By rights, twentieth century physicists shouldn't have had the priviledge of studying this theory. By rights, string theory shouldn't have been invented until our knowledge of some of the areas that are prerequisite for string theory had developed to the point that it was possible for us to have the right concept of what it was all about.

We need twenty-first century mathematics?

Probably. What should have happened, by rights, is that the correct mathematical structures should have been developed in the twenty-first or twenty-second century, and then finally physicists should have invented string theory as a physical theory that is made possible by those structures. If that had happened, then the first physicists working with string theory would have known what they were doing perhaps, just like Einstein knew what he was doing when he invented general relativity. That would have perhaps been a normal way for things to happen but it wouldn't have given twentieth century physicists the chance to work on this fascinating theory. As it is we have had the stroke of good luck that string theory was invented in a sense without human beings on planet Earth really deserving it. But anyway we have had this stroke of good luck and we are trying to make the best of it. But we are paying the price for the fact that we didn't come by this thing in the usual way.

Just one further point, as far as the higher dimensions are concerned. How many different ways are there for the higher dimensions to curl up on themselves?

Physicists working on the problem have imagined many conceivable ways for the higher dimensions to curl up and very probably there are quite a few others that haven't been conceived of yet. In fact, very probably the whole process will turn out to be more subtle than we have imagined so far.

Are these all contending theories or are we going to be able to isolate a particular way of curling up the higher dimensions?

I imagine that we would need to understand better what kind of theory it is that we are dealing with before we would successfully find the right way of doing that.

It seems at the moment that this must be a rather major obstacle to further progress in the theory — not knowing how the higher dimensions curl up.

We would be much happier if we understood how the higher dimensions curl up and therefore what the vacuum state of the theory is.

Do you need to know this before you can calculate any of the fine details of particle masses or charges or such like?

That's right. Then we get the pay-off of calculating the masses, the lifetimes, the interactions and so on of the elementary particles. Regretably I don't think it is very likely at present that one would be able to do this successfully in the near future. I think there is probably quite a bit about string theory that we would have to learn before being able to do it. That's only my conjecture. There have been a lot of ideas about how the extra dimensions could curl up and a lot of people tinkering around with variations along the known themes and interesting ideas continue to turn up. I just heard about a new one yesterday as a matter of fact.

So what is the main thrust of present investigation, given that this central problem is going to remain unsolved for a while?

A lot of physicists are working in different ways on what I would say was the essential intellectual question of what string theory really means. What is the analogue of the symmetry principles on which other physical theories are based?

Some people might think that it's a little bit unsatisfactory that there is a small army of physicists devoting their attention to a theory which quite likely won't make contact with experiment maybe for another generation. Do you think that it is justified for this tremendous attention to be given to the subject?

I can only speak for myself. I think that it is a tremendous stroke of good luck to be working in physics during the epoch in which string theory is developing. I personally believe that in future centuries people will look back and say that this was one of the great times to do physics.

Looking back at the past fifty years, do you think that the approach to fundamental physics is changing at all? Are the sort of techniques and the sort of philosophy employed in the superstring programme fundamentally different from those of our forefathers?

The style of physics has changed a lot over the decades for many reasons, most of all because of progress that's taken theoretical physics into new domains. Some fifty years ago quantum field theory was a complete mish-mash, then in the course of time progress was made. By understanding it better, it was successfully brought into new domains, was made to encompass new interactions, and it was given more of a geometrical basis, which doesn't quite match that of general relativity but at least is a worthy contender.

Physics progressed in a way which made it possible to be more ambitious about what one regarded as a satisfactory answer to a physical question. It's good to bear in mind that in the nineteenth century physicists didn't even have the aspiration to explain why glass is transparent or why grass is green, why ice melts at the temperature it does, and so on. Those questions were not part of physics in the nineteenth century, and physicists didn't even dream of being able to answer questions like those. They had more modest aspirations. Given certain measurements about how flexible a material was they hoped to be able to calculate the outcomes of certain other experiments, but to predict the whole kitcat and caboodle from basic equations about electrons and nuclei as became possible in the twentieth century, this wasn't even a dream in the nineteenth century.

The progress of physics has always been such that the level of understanding for which one generation aims wasn't even dreamed of a generation or two earlier. Twenty years ago elementary particle physics was again a mish-mash with a vast collection of elementary particles discovered, and it was not at all clear what was the right framework for describing them. A satisfactory framework for the known forces, except gravity, emerged in the period around 1970. This brought order to the chaos of the elementary particle world and created a completely new environment for thinking about elementary particle physics. So if nowadays we have different kinds of questions on the table and we address them with different methods, it's largely because of the progress that was made in that period and, of course, in earlier periods.

Stephen Hawking, although he wasn't referring to the superstring theory but to similar attempts to tackle these very fundamental issues, has claimed that the end is in sight for theoretical physics. Do you think if the superstring programme is pushed to a successful conclusion, maybe in fifty years time, then it will be the culmination of theoretical physics? Will it wrap the subject up once and for all?

The first real try at quantum mechanics in the atomic domain was Bohr's model of the hydrogen atom in 1914. After the Bohr atom it was pretty clear that there was something true about quantum mechanics in the atomic domain, but it wasn't clear what. There was a period of confusion and it wasn't at all clear what the scope of quantum mechanics was going to be. The scope of quantum mechanics turned out to be a lot more radical than anyone had imagined. Only after the Schrödinger equation in 1925 did it begin to emerge what quantum mechanics did and what it didn't do, and just how radical was it's impact on human thought.

I think that we are in a similar period with string theory. I think that even most people who are enthusiastic about string theory tend to underestimate how radical it will prove to be in its impact on how we understand physical law. We are uncovering part of the structure, but we haven't got to the nub of things yet. Again, as with quantum mechanics, I think that without coming to the nub of what string theory really is, it's hard to foresee what theoretical physics is going to be like after that. I think that theoretical physics will be on a plane that we can hardly imagine today. What the problems will be in that epoch I wouldn't care to speculate about.

4

Michael Green

Michael Green is Professor in the Department of Physics at Queen Mary College, London. As one of the originators of modern string theory, it was largely through his work in collaboration with John Schwarz that the subject has been propelled to prominence.

Can we start by going back to the early days of string theory when it was still something of a backwater subject. Could you say something about how you first became involved and what you were trying to do in those days?

Well string theory has a very curious history because it was originally invented, or grew out of interest in a rather different field of physics from the one in which it is currently generating interest. In those days, the strings were meant to describe the hadrons – the strongly interacting particles like protons and neutrons. Roughly speaking, you could think of these particles as being made up of quarks which were tied together by string. The subject grew out of intense interest in strong interaction physics, the physics of these particles, in the late sixties, most notably following work by Gabrielle Veneziano, an Italian physicist. At that time I was doing research for my Ph.D thesis, and was immediately struck by these interesting new ideas. This was partly because they were so very different from the conventional ideas based on quantum field theory which had notably failed to accommodate the sort of physics involved.

If I may say so, it seems a strange sort of model to be applying to hadron physics, to the physics of these strongly interacting particles, imagining that they have got little strings inside of them. Didn't it seem a bit bizarre? Did you really think that it was going to describe reality?

It wasn't that the particles had strings inside them. In those days the thinking was that the particles themselves were string-like. So, for example, a pi-meson, one of the most basic strongly interacting particles, could be thought of, roughly speaking, as a quark and an antiquark tied together by string. The fact that these quarks are tied together by string would be part of the reasoning for why quarks could not be seen separately.

A bit like a dumbbell, and the whole thing could be whizzing round, I suppose?

Yes, that's right. In fact, this is very much the picture that has now emerged from the theory called quantum chromodynamics, which is the modern theory of the strong interactions. In a certain sense that can also be thought of in terms of this older string picture.

So there are still vestiges of string-like qualities in the more modern description of hadrons?

Yes, I think that's a good way of expressing it.

What about the case of a particle like a proton which has three quarks inside. Wouldn't you need to have three strings connecting them together?

That's right, and it was because of issues like that, as well as other very severe technical problems, that that particular application of string theory died eventually.

The history, in fact, was even more bizarre than I have indicated because Veneziano's original suggestion was essentially just a guess as to what would happen when strongly interacting particles collided. He didn't have a string picture in mind at that time, but he made this very inspired guess which triggered off a lot of other people's research into the structure of the model he proposed. Eventually, after about two or three years, it was realized that this structure would emerge if one thought of particles as being string-like. So it took a bit of time before it was even realized that his guess was based on a picture in which particles are strings.

Well clearly, as you say, that particular application of string theory didn't get too far, although it had some promising signs. What happened? Did it just fade out of interest?

Its early history – that means the early seventies – coincided with the revolution in the understanding of the weak force in the context of a unified theory of the weak and electromagnetic interactions. Also there was enormous progress in understanding the strong force in rather conventional language – in the language of quantum field theory, which is the basic tool of the subject. Because of the enormous theoretical progress at that time in understanding the theories of those forces, as well as an explosion in experimental successes verifying these theories, I think that most people's attention was caught by these more conventional topics rather than by string theory. But that was also the time in which a group of people were mesmerized by string theory. String theory is, once you have learnt it, so captivating, so elegant, that it's very difficult to put it out of your mind. I think that motivated almost all the people involved more than any direct application of the theory to any particular branch of physics.

Why is this? What is the secret of the success of string theory? Why is it so captivating?

It's partly because the theory contains the sorts of structures that we are familiar with in what we normally think of as being beautiful quantum theories. For example, theoretical physicists love gauge theories, these are theories like electrodynamics and the theory of the strong force, and indeed Einstein's theory of gravity. These are all considered to be very elegant theories because they embody a type of symmetry known as gauge symmetry which allows the theory to be consistent in a way that it wouldn't be otherwise.

We're talking here about certain mathematical symmetry properties which are apparent to theoretical physicists, but not to members of the public. They nevertheless bring joy to the hearts of theoretical physicists.

In a sense that's right, yes. It's very difficult to describe extended objects like strings in a way which is consistent with Einstein's special theory of relativity. At first sight, the theories suffer from terrible problems that one would think would render them inconsistent.

What sort of problems?

The most obvious is the fact that, superficially, they appear to describe strings which have unphysical modes of vibration. They vibrate not only in space but in time. They wriggle in a direction which doesn't make sense, in a time-like direction.

The captivating thing about the early string theories was that although they had this apparent problem in fact they avoid it in a way that is reminiscent of the way in which the analogous problem is avoided in Maxwell's theory of electromagnetism. But they avoid it in an infinitely more subtle way, because the problem is infinitely bigger. The fact that it can be avoided at all is remarkable.

How do they do it?

In order to avoid these apparent inconsistencies, the theory only makes sense if certain conditions hold, namely if the string is moving through space and time where the dimension of space is fixed at a certain value. In the original string theory, space had to have twenty-five dimensions, spacetime has twenty-six dimensions. In the later string theories the theories worked only if space had nine, and spacetime ten, dimensions.

What did people make of that in the early days?

It was considered to be a disaster, because we live in what appears to be three space dimensions plus one of time. So this counted very much against the string theory in those days.

There was another problem, which I think was much more severe, because this really was an inconsistency. Namely, these theories contained particles that travelled faster than the speed of light, so-called tachyons. If you're not worried about quantum mechanics, the existence of particles travelling faster than the speed of light is something that one might contemplate, but in a quantum theory it doesn't seem to be possible to make sense of such particles.

Some people are familiar with the theory of relativity, which also says that if you have things travelling faster than light it might be bad news from the point of view of causality.

Yes, but it may not be altogether bad news. One might be able to circumvent that in a classical theory, simply by not allowing contact between slower-than-light and faster-than-light systems. The real problem arises in a quantum mechanical theory, because the notion of the lowest energy state of the system then ceases to make sense if there are tachyons. The state that we think of as being empty space – the vacuum – would be unstable, because it would be able to decay into these particles. In other words, the vacuum would just explode into an infinite number of tachyons. So we don't know how to make sense of a theory which contains such particles.

So this was the state of affairs round about the mid-seventies, is that right?

Yes, that's right. Around the mid-seventies, many of the theorists working on this subject were captivated by the advances in the more conventional theories that had grown up in those days, as I explained before. One can track the activities of this population of physicists fairly accurately, and broadly speaking they went into two types of research. At that time there was just developing an understanding of issues in certain conventional gauge theories that went mathematically far beyond what had been understood before. It was understood, for example, how these theories contain magnetic monopoles – magnetic charges – in a very subtle way, a way which hadn't been anticipated earlier. The second novel line of theoretical research was supersymmetry.

What exactly is supersymmetry?

Symmetry principles play a very important role in the development of elementary particle physics, principally because they reveal patterns of properties that relate particles which are seemingly different, and once one has seen these patterns one gets a clue as to the structure of the underlying forces. A good example of the use of symmetry in science is that of nineteenth century chemistry. In the last century it was understood by Mendeleev that one could arrange the chemical elements into groups with certain properties in common.

The famous periodic table.

That's right. The periodic table has groupings of elements which fit together in certain patterns, and then it was understood that these many dozens of elements could be grouped in this way because of the fact that they were made of atoms. The patterns were seen to emerge from an understanding of a single phenomenon, namely the electric force which holds the electrons in orbit around the atoms.

The hope in particle physics is that by grouping particles together according to their properties, one might get a clue as to what the underlying forces are.

The study of the forces between these particles has developed enormously and we now have an understanding of the strong force and the electroweak force, within which the particles group together in a certain way. However, as of the mid-seventies, the groupings fell into two distinct classes. Particles have a property called spin, which is a sort of angular momentum – in a simple-minded fashion you can think of these particles as simply spinning around an axis – but in quantum mechanics this spin only occurs in discrete units. Particles in which the discrete unit is an integer are called bosons. Particles for which this discrete unit of angular momentum is a half-integer are called fermions.

Up to the mid-seventies, although groupings of fermions and of bosons had separately been understood, it wasn't understood in what sense the fermions and bosons might be related to each other by some sort of symmetry. In other words, fermions and bosons seemed to be completely disjoint, whereas if we want a really fundamental understanding of the origin of all the particles in terms of a single principle, we would really like to understand the relationships between these two groups of particles.

Now supersymmetry, which emerged in theories in the mid-seventies, is a symmetry which relates fermions to bosons, so that if supersymmetry is a symmetry of physical laws, then these two apparently different groupings of particles are, in fact, different aspects of the same object.

And the 'super' in superstrings arises from the supersymmetry that has been built in. Is that right?

That's absolutely right.

What effect did applying supersymmetry to the old string theory have?

Amazingly enough, taking one version of the old string theories, and modifying it so that it is supersymmetric, instantly gets rid of the problem of particles moving faster than the speed of light. These particles no longer occur in the theory, and the theory appears at that point to be consistent in a way that it hadn't been earlier.

At that stage was it generally appreciated that you were now onto something really rather more exciting?

Well, there had been a gap in the development of string theory from around 1976 to 1979, during which time just about everybody had stopped doing research in the subject. Now this is rather curious because, in fact, in 1976, just before this gap, there had been a paper published (by Jöel Scherk, Ferdinand Gliozzi and David Olive) which had suggested the possibility that the supersymmetric modification of string theory would be interesting, but it wasn't pursued, and the subject more or less died.

Then I started to work with John Schwarz in Caltech in around 1979, and we pursued the idea of making string theories supersymmetric. We were struck by the fact that the theory looked consistent at that point. I must admit that at that time very few of our colleagues were interested, again principally because there had been important developments in another field which looked very interesting – the field of supergravity, that is, applying supersymmetry to gravitation. And it didn't seem to the people involved that string theory was worth the investment of effort needed to really understand it.

What induced you to think of marrying together supersymmetry and string theory? I can understand you were obviously interested in strings anyway, but was it a very obvious thing to do, to try and make it supersymmetric?

I think it was in a way. I mean, everybody was making everything supersymmetric in those days! Supersymmetry seemed like a beautiful new idea in physics because it really was the last link, in a general sense, in the unification of different types of particles. There had never before been an understanding of how symmetry could relate particles of different spin, and supersymmetry provided this link in the understanding. So from a purely theoretical point of view, it seemed that something like supersymmetry was essential to any theory aimed at unifying the particles, despite the fact that there is not yet any experimental evidence for such a symmetry in nature.

When you embarked upon it did you expect to get dramatic results, or did it surprise you when things seemed to be working out so well?

I think for the first couple of years that we were investigating these theories, we were still just mesmerized by the fact that by studying them in more and more detail, one came up with more and more ways in which these theories were consistent.

There is a very clear date at which we became convinced that we were on to something quite important, and that was in late 1981. We showed that a certain quantum calculation of one of these superstring theories gave a result that was not nonsense. I put it that way simply because these are theories containing gravity and all quantum theories of gravity up to that point gave nonsense in the sense that I'm referring to now.

What sort of nonsense are we talking about?

Well, if you try to calculate the probability of two particles scattering in such a theory you find you always get an infinite answer. That's what I mean by nonsense.

And in your calculations you found that you could get some finite answers?

We discovered that, at least in the simplest approximation to one of the theories that we were looking at – this is a theory which only contains closed strings – the theory was actually finite. Now that was quite startling because this was, after all, a theory containing gravity. Conventional theories of gravity based on Einstein's general theory of relativity give terrible problems, even at this lowest order of approximation. So we were very struck at that point with the fact that we were probably onto something which was exceptionally interesting.

Let me get this clear. The finite results came from a calculation which was only an approximation. Could you say that the theory was definitely consistent?

No. Certainly what we were talking about then and what is still being talked about now is finiteness within the context of an approximation to the complete theory.

How did people respond to that initial finding?

Well, very few people paid attention. It was certainly a result which should have attracted attention (from a certain group of people anyway) since there was at that point a fairly large bandwagon of people working on supergravity, which was also an attempt to formulate a consistent quantum mechanics containing gravity. But people virtually ignored us. There were one or two people who were certainly struck by our result, in particular Ed Witten. In fact, he and Louis Alvarez-Gaumé went on to show that it was very plausible that the same string theory was not only finite but was free of the other problem that plagues theories of quantum gravity, namely anomalies. So the fact that they were motivated to look at that makes it clear that they were very interested at that point. But they were exceptions, I think. Most people just simply felt string theory was too far away from conventional quantum field theories.

Now, as you said, the theory that you showed to be finite, at least at this first order of approximation, concerned closed strings. But closed strings were, at that time, considered not to be a useful concept for a theory that was going to tell us something about the real world. Is that correct?

Yes. To be fair to the people who ignored what we were doing, it is certainly true that that particular theory, which is a theory only of closed strings, didn't look as though it could possibly make contact with the laws of physics, apart from the force of gravity. It just didn't have enough structure in addition to the gravitational force to make it plausible that one could describe the other forces using that theory, although even that wasn't totally clear. In fact, it's rather curious because the most recently discovered kinds of string theories are also theories which only contain closed strings. They are generalizations, if you like, of the theory that we were looking at then. These generalizations do indeed have a lot more structure in them, and these newer types of theories, called heterotic string theories, are the kinds of theories that most people think might explain the other forces.

Perhaps we ought to discuss a little bit here about the different types of superstring theory that there are. This sounds like bad news for a start, because, presumably, if one is searching for a Theory of Everything, one would hope that it would be unique. Just how many different string theories are there?

Well that depends on how you count. In a certain way of counting there are four or five, but really that may not be the right way of counting, and in a different way of counting one could argue that there are many thousands at the moment.

The reason that I say that it depends on how you count, is because these many thousands are actually rather naturally thought of as being different versions of the four or five.

I would like to point out that although it's bad news at the moment that there appear to be so many versions, this subject is really in its infancy and every time one studies some other aspect of string theory, one discovers that there are new apparent difficulties that have to be overcome and that in order to overcome them the theory has to be very much more special than you first thought.

You're saying that these many contending versions of string theory are still, in a certain sense, incomplete, and when there's a full understanding of how they work it may be that some of them will just self-destruct, that they won't be able to give a fully consistent account of the world?

That's my hunch and it's based on the history of the subject in earlier days. For example, around 1982 when we were getting excited by the fact that these theories appeared to be finite, we thought that *all* superstring theories would be finite. At that time we thought we had the option of introducing into these theories arbitrary symmetries for the other forces in nature, apart from gravity. The history of the unification of the forces in the late seventies was that people felt that there should be something they call the grand unified theory or grand unified symmetry, some huge symmetry which was a set of mathematical relationships that link all the particles we see in nature, and would encompass all the forces apart from gravity in a single scheme.

Now in the traditional grand unified approach, one selected a particular symmetry on the basis of experimental information rather than for any theoretical reasons. There was no theoretical reason in those days for choosing one particular sort of symmetry relating the particles rather than some other symmetry, and we felt, just because we were influenced very much by the way grand unified theories were meant to work, that we could also introduce an arbitrary symmetry relating all the nongravitational bits of the superstring theory. Any one of these would be theoretically as good as any other and we would likewise have to choose one of them on experimental grounds eventually. But that wasn't what happened. So here we have an example where it appeared that there were an infinite set of possible theories, with different symmetries, but in fact we then discovered that there were only a very limited number of such theories which were really consistent.

So what's the difference between a heterotic string and the kinds of strings that you envisaged back in 1982?

Well, the heterotic theories are rather curious beasts. They can be thought of as combinations of the oldest kind of string theory, the original so-called bosonic string theory, on the one hand, and superstring theory on the other. So the heterotic string combines the string theory which works in twenty-six-dimensional spacetime with one which works in ten dimensions! That, of course, doesn't make sense. You can't have a different number of spacetime dimensions for the same string. What actually happens is that ten of the twenty-six dimensions are ordinary spacetime dimensions, so that the string is wriggling in ten-dimensional space-time. In addition there are sixteen so-called *internal* dimensions. These lead to extra structure in the theory that ought to describe the other forces, the forces other than gravity. So there's a rather geometrical picture for where these other forces come from. They come from the fact that twenty-six minus ten is sixteen! The sixteen mismatching dimensions are responsible for certain symmetries of the theory. These symmetries go under the names $SO(32)$ and $E_8 \times E_8$, which are mathematical names for the relationships between the particles in the theory. $SO(32)$ and $E_8 \times E_8$, are mathematical *symmetry groups* which are naturally associated, in the heterotic theory, with the mismatch of sixteen dimensions between the bosonic string theory and superstring theory.

So is it correct that the sixteen extra dimensions in the heterotic theories that are referred to are somehow related to the non-gravitational forces?

Yes, the difference between the heterotic superstring theories and the other string theories that have some chance of making contact with physics – namely some open string theories – is that in the open string theories the charges associated with the forces – such as the electric charge, the charges responsible for the strong force and so on – these charges reside at the end points of the string. In the heterotic theories the strings don't have end points, they are closed strings, and one can think of the charges as being smeared out over the strings. That's the chief physical distinction between the two kinds of theories.

How would you envisage an electron in terms of closed strings? An electron is a charged particle; should one envisage the charge as being smeared along the string?

In a closed string theory of the heterotic type, something like that is the right picture. The string can vibrate in any of an infinite number of different harmonics, and any one particular frequency of vibration corresponds to a particle or a set of particles. Now the particles which we actually observe in nature, things like the electron or the quarks or the photon or other particles, *all* of the particles that we observe would actually be the lowest possible mode of vibration of the string, in a certain sense the mode in which the string isn't vibrating at all.

You are saying that the particles which exist in nature all correspond to a nonvibrating mode of the string. How is it then that this string, in its nonvibrating mode, can lead to many different types of particles?

Well, there's more to string theory than the simple picture of a string vibrating in space. In the earliest string theory that simple picture was correct but that theory didn't contain the particles that we know, and it had other inconsistencies as well. In the more realistic theories, the superstring theories, there is extra structure besides the fact that the string can vibrate in space. There are charges like the electric charge and the weak charge and so on, which reside on the string, and it's the nature of these charges which distinguishes different particles like electrons, neutrinos, quarks, etc. So any given type of vibration of the string in space corresponds to a set of particles, not just to one particle. In particular the ground state of the string, the state in which it is not vibrating, doesn't just describe a single particle, it describes a whole bunch of particles, and these are the particles that we are supposed to see in the laboratory.

If you're having to rely on the idea of charges that can be distributed in different ways to explain this multiplicity of particles, doesn't that remove one of the advantages of string theory, which is that it is supposed to explain everything, like electric charge, in terms of geometry?

My description of the theory in that language was just meant to be an intuitive way of understanding what emerges. One doesn't have an arbitrary distribution of such charges. The way in which these charges arise and how they are distributed is specified extremely clearly by the theory. Only theories with very specific kinds of charges distributed in particular ways can possibly be consistent with quantum mechanics. One cannot talk randomly about strings with charges distributed in any way one wants. Only specific kinds of theories are consistent.

How many different types of charges are there?

The theories that were discussed originally contain the symmetries $SO(32)$ and $E_8 \times E_8$. These have 16 different kinds of charges, and 496 gauge particles like the photon, which transmit the forces exerted by these charges. One doesn't have the freedom to alter things at will in these kinds of theories. This feature distinguishes string theory from earlier kinds of theories based on point particles.

So the kinds of charges that you're talking about don't necessarily correspond to the kind of charges that most people understand, like electric charge.

Well, they contain the concept of electric charge, as well as weak and strong charges, within them. In searching for some unifying description one tries to build up a picture in which the different types of charges are unified within a bigger structure, and this bigger structure will describe new particles with many other charges besides the ones which are seen directly in the laboratory. Some of these we may eventually see, and some may be particles with masses so enormous that we may never see them. The string theories being talked about two or three years ago had this enormous number of 496 gauge particles. This included the ones we see and many others that have not been seen.

It's been quoted that there was an occasion when the number 496 occurred in a calculation of yours, and when it popped up you suddenly realized you were onto something important. Can you tell me about that?

Well, in the summer of 1984, John Schwarz and I got around to asking the question whether the string theories that might be of interest to physics, were consistent in the sense of whether or not they had anomalies. Now it was already strongly indicated from the work of 1981 and 1982 that the closed string theory was consistent but that didn't seem to be of direct relevance to physics. What we expected was that open string theories, the ones that might be of interest to physics, would avoid the problem of anomalies completely. Now this expectation was not based on anything more than wishful thinking. I think most other people expected that string theory would *always* have the anomaly problem because it seemed on very general grounds that anomalies were something which string theory couldn't possibly get around. We, on the other hand, being very optimistic, felt that string theory was so magical that it would always get around the anomaly problem, and we were amazed to discover that the truth actually lies somewhere in between. It turns out that almost all string theories are indeed sick, they almost all possess anomalies. But amongst the theories that we were looking at was a single unique theory which avoided this problem. When we discovered that we were very intrigued, but the way in which we had discovered it left open the possibility that it was a coincidence, since we had only looked at one particular type of anomaly out of many. Finally, one day we did a rather more sophisticated calculation in which we looked at all possible anomalies at once, and for that calculation to work there had to be a miraculous cancellation between many peculiar looking numbers so that the answer of adding these numbers together had to be 496. And indeed that's what happened!

Can you tell me what the present verdict is on whether these theories are finite or not because I gather there is still some argument over this?

Well, the situation has still not been clarified but I think there is a consensus about what people imagine is going to happen. String theory has always been considered, and is still considered, in terms of an approximation procedure. We have never yet solved any string theory exactly. In this procedure one makes successively better and better approximations and one has to, at each stage, ask whether the latest approximation still gives a finite answer, because trouble might arise at any order in this procedure.

Originally we looked at the lowest order approximation, the simplest approximation to look at, and that was finite. This was immediately striking all by itself, because even at that level, no previous quantum theory of gravity had given a sensible answer. Nobody has yet managed to show that all possible orders in this approximation procedure for these theories give sensible answers. However, the way string theory works makes it very plausible that once one has looked at the lowest order of approximation and found that is consistent, it will be consistent in this sense to *all* orders in the approximation procedure. So although there is certainly an outstanding issue and many people are trying hard to figure out why these theories should be consistent to all orders, I think people generally believe that they *will* be – at least those which are consistent to the lowest order. However, it's a very interesting study nevertheless. By trying to solve this problem to all orders in the approximation procedure, one is discovering properties of the theory which go beyond any particular step in the approximation. That's actually one of the main themes in the current phase of string theory research.

Returning to the historical narrative, there you were in 1982 suddenly discovering that you could get sensible answers in gravitational calculations, and yet you had embarked upon this great enterprise believing that you were presumably describing something like strong interaction physics.

By then that wasn't the case. As soon as the theory was made supersymmetric, in other words, as soon as we had the structure of the superstring theory, it was obvious that it somehow bore a strong relationship to supergravity theories.

Clearly you already anticipated that a description of gravitation would come out of these theories.

We certainly knew the theory contained supergravity in some form. Supergravity is, in fact, contained as an approximation to superstring theories. It is an approximation which by itself is not consistent but there is a sense in which supergravity is a piece of the superstring theory.

So this went beyond the supergravity ideas, which were very popular at that time, to develop something new.

Right. String theory is radically different from any of the earlier theories simply because all earlier theories – from Maxwell's theory of electrodynamics to general relativity and supergravity – contain particles, such as the photon, graviton, quarks and other particles, which are point-like objects, they have no internal structure. String theory differs from this in that the constituents of string theories are extended objects – strings. Although that may sound like a very mundane difference, it actually makes an *enormous* difference to the structure of the theory.

Is it easy to see how it makes a difference?

Well, one can give a hand-waving argument for why it makes such a difference. In quantum mechanics it is very awkward, if not impossible, to deal with point-like objects. One way of describing quantum mechanics is in terms of the so-called uncertainty principle and by using the uncertainty principle, it's easy to argue that the shorter the distance scale that you're trying to describe, the more uncertainty there is in the energy of what you're trying to describe. Now in a theory of gravity this means that when you try to describe things at incredibly short distances (and by short I mean *unbelievably* short compared to the size, say, of even the proton), the fluctuation in energy of what you are looking at might be big enough to make a little black hole. So if we are contemplating making an observation on small enough distance scales (this scale is called the Planck distance, it is 10^{-33} centimetres), we are forced to think of even empty space as consisting of an infinite sea of fluctuating black holes, coming and going in very short times. This of course radically alters our notions of what we mean by space, and it's a disaster because we don't really understand what's happening any more. The whole notion of space itself as being made of points probably no longer makes sense.

But is there no way one can have a point-like object moving in such a spacetime background?

The string is an incredibly short object. On average its length is the Planck length, i.e. twenty powers of ten smaller than the size of a proton. So that for many purposes the fact that it's a string doesn't really matter. You wouldn't notice that it was an extended string if you weren't looking incredibly carefully.

You mean it behaves like a point particle apart from at these very short distances and very high energies?

That's right. And these are distances that we would never hope to be able to measure directly in the laboratory anyway. However, these are precisely the distances on which all the problems of quantum gravity appear, and it's just at those scales that the string theory begins to differ radically from Einstein's theory, or indeed any other of the earlier theories.

Would it be an incorrect image to suppose that if a little string, which might be closed into a loop, is looked at on a coarse scale, it appears just like a particle, but that if one could go to a very fine level of detail what you would see is all sorts of wriggling motion, and it's that wriggling motion which modifies the high energy behaviour?

There is a certain sense in which one can use that picture and that is indeed the way that most of us think of the theory at the present time. But actually the theory is probably much deeper than that, because just as one is looking at those scales on which one would see the wriggles, so to speak, one is then looking at the scales in which the whole structure of space and time has to be modified. Therefore it may not be correct even to think of this thing as moving through what we normally think of as continuous space and time.

The strings are actually wriggling around in a shifting spacetime background?

In a theory of gravity you can't really separate the structure of space and time from the particles which are associated with the force of gravity, and since we now have enlarged what we mean by gravity, so that Einstein's theory of gravity is only one small piece of this theory, then we also have to enlarge greatly what we mean by space and time.

Are you saying that space and time are in some sense built out of strings rather than the strings inhabiting space and time?

Yes. The notion of a string is inseparable from the space and time in which it's moving, and therefore if one has radically modified one's notion of the particle responsible for gravity, so that now it's string-like, one is also forced to abandon at some level the conventional notions of the structure of space and time. When I say at some level, the level I'm talking of is at these incredibly short scales associated with the Planck distance.

Let me see if I understand this correctly. For almost all purposes we can forget about the shifting, nebulous nature of space and time on the small scale, and envisage particles inhabiting ordinary background space and time in the usual way. But if we look on a finer level of detail we begin to see these strings, and space and time and the strings become interwoven in some very subtle way.

That's right. It's a subtle way that hasn't really been understood properly yet. A lot of the present research is focused on trying to understand precisely how that works.

In 1984 the interesting superstring theories only made sense if spacetime had ten dimensions. How should we envisage the relationship between the ten-dimensional spacetime background and the space and time of our perceptions, which, of course, has only four dimensions?

Well, obviously the extra dimensions have to be different somehow because otherwise we would notice them. What one has to get used to is the idea that in any theory containing gravity, the theory itself defines the structure of space. One has to get used to the notion that space can be curved, that dimensions can curl up and become, in some sense, very small.

It's a difficult idea to grasp but one can think very loosely of a simple analogy where there seems to be one dimension less than there really is. Think about a hosepipe for example. A hosepipe is a two-dimensional surface, it's a long object which has one circular dimension. Now if you don't look too closely at the hosepipe, you might think it was a one-dimensional object, just a line in fact. But if you look closely you then realize that there is another very small dimension – the hosepipe is actually a narrow tube.

In a generalization of that example, there could be several extra dimensions which are so tightly curled up that you wouldn't notice them unless you could look, somehow, with extremely fine resolution.

This means that each point of space, or what we thought was just a point in space, is actually a higher dimensional curled-up object.

Correct.

It might appear a bit mysterious that in a theory that starts out with ten dimensions, we end up with four and that six curl up. Why six?

Well, certainly we don't really understand that at the present time. I think we have only just begun to probe the tip of the iceberg in our mathematical understanding of these theories. We've realized almost by a series of accidents that the theories are very special, by doing rather mundane calculations and getting rather extraordinary indications. But the full structure of the theory is not understood, and that sort of question is one for which I don't think we can give a proper answer until the theory is reformulated in a way which makes it more complete mathematically. For example, recent developments have thrown up variants of superstring theories which work directly in four spacetime dimensions – the extra dimensions are in some sense automatically curled up.

Assuming that one day you come to understand why it's six dimensions that curl up, do you think you are also going to be able to understand the way in which they curl up? There must be many ways of folding six dimensions onto themselves, many different topologies.

I personally believe that we will but, in fact, it's a matter of dispute as to whether that is the sort of thing that we should *ever* be able to understand. It's logically conceivable that there are many possible ways in which it could happen and it's just an accident that we happen to live in the universe in which the extra dimensions are folded up in this particular way.

Could it be that if they folded up in some other way, then conditions simply wouldn't be suitable for life?

Well that is one possible argument, but it's not the sort of argument that I find attractive.

Do we have a problem in that, as you said earlier, the structure of space and time become sort of frothy or foamy, or whatever is one's favourite description, on a very small length scale, and yet you wish to formulate the string theory by starting with, presumably, a conventional background space and time?

Yes. Obviously that isn't the fully correct way of doing things. That can only, at best, be some sort of approximation to the real world, but at the moment it's the best that's been done. However, what we've seen is that, even at that level, the theory only makes sense for a very restricted class of symmetries in the theory, and that's already very interesting. Indeed, that restricted class is particularly interesting because at least one of the theoretically possible symmetries bears a very striking resemblance to the sorts of symmetries that had earlier been suggested on purely experimental grounds as possible symmetries for describing the particles we see in experiments.

I think it's true to say that a great deal of the present excitement in superstring theories is due to the fact that one of the possible kinds of theories, the $E_8 \times E_8$ theory, involves what are called exceptional groups. These are very particular mathematical symmetries that play a very special role in mathematics and have been conjectured, for that reason, to play a very special role in physics. Well, now in the superstring theory we finally have a theoretical reason why they arise in physics, and that is what, I think, excited a lot of people.

You're saying that nature has spotted a rather exceptional piece of mathematics, known as the exceptional group, and is making use of it somehow?

That's right, and I think that's very appealing to theoretical physicists. The way in which these theories are meant to make contact with physics is somewhat difficult to calculate at the present time, because the sorts of calculations which are easy to make are calculations of things that would be measurable only if one could explore incredibly small distance scales, or equivalently high energies, and we can't do that directly in any laboratory. So what one has to do is understand how to extrapolate from the physics at very short distances, to see what it predicts for the sorts of distances that can be measured in accelerator laboratories on Earth. That kind of extrapolation is very difficult to perform.

Nevertheless, what has been done so far is very intriguing and exciting because there are all sorts of very severe theoretical limitations as to what *can* happen. So, for example, although we can't prove that the extra dimensions really are curled up and are unnoticeably small, if we *assume* that the equations of a theory will eventually predict that they do curl up, then all sorts of quasi-predictions fall out immediately. It's very intriguing that *if* we assume that extra dimensions – the ones we don't want – are in fact very small, then there is a clear path by which the sorts of symmetries that are seen directly in the laboratory in experimental work in high energy physics could, in fact, be predicted by the theory.

You mentioned that more recent developments have opened up the possibility of developing string theories not only in ten dimensions but now in other dimensions.

In 1984 we had a more-or-less unique specification of what the theory should look like in the approximation that space and time had ten dimensions. In that situation one had the choice of only two possible theories: one with $SO(32)$ and one with $E_8 \times E_8$ as the symmetries of the particles. Of course, ten dimensions is not the dimension of spacetime in which we live, and what people realized rapidly was that if the six extra dimensions were curled up and very small, then it was quite plausible that these heterotic theories would give rise to sensible four-spacetime dimensional physics. Now even at that stage it was clear that that could happen in many different ways. One had a starting theory that was more-or-less unique, but then there were many different possible solutions to that theory which worked in four spacetime dimensions. We didn't know how to select which one of the solutions of the theory was the correct solution, even though we might have a unique, or almost unique, specification of what the starting theory was.

Now people have discovered ways of constructing new kinds of solutions which work directly in four dimensions. In other words, they never have to pass through the stage of ten dimensions. These are the different versions I talked about earlier and it's wrong to call these different theories. They can be viewed as different kinds of solutions to the same theory as the ten-dimensional ones. So we are in a situation where we have a very wide variety of solutions to rather few theories.

There is an analogy here. Imagine that you were shown specimens of ice, water and steam. It may take you a while to recognize that these are actually different phases of the same substance and that the underlying laws of physics governing the microscopic properties of these objects were the same. It is just a matter of the conditions of the substance, the conditions under which you are looking, which distinguishes these three different phases of water.

The situation in superstring theory is roughly similar. There are a very large number of different phases corresponding to these different solutions to the theory, and we have yet to discern the underlying structure. This is, in fact, the main goal of much of the recent work, trying to find a more fundamental setting for superstring theory so that we will then have a set of equations with approximate solutions which are these many different 'theories' that we have at the moment. Hopefully, we will then be able to figure out which, if any, of these solutions describes observed physics.

One thing slightly puzzles me about formulating the theory in four dimensions directly. I thought that the anomaly cancellations would only work if the theory was formulated in ten dimensions.

Well, as I have said, all these different 'theories' are really different phases of the same underlying theory. The anomaly cancellation works for all of them. In this picture where one formulates the theory in terms of successively better and better approximations, one is thinking of particles which are string-like moving through space and time which is more-or-less a classical space and time as we have known it before. But string theory is really much deeper than that. As I have already explained, it ought really to alter what we mean by space and time as well as altering what we mean by particles. Now that aspect of string theory, this really deep aspect whereby the spacetime in which the string is moving is itself altered by strings, is not really contained within the present formulation of string theory. What we really need is a fundamental new idea about the formulation of string theory which will embody this principle. Then, of course, one ought to find that the approximations we are using come from this more fundamental picture. But at the same time one will probably, at that point, understand the difference between the many kinds of different phases or different solutions of the theory. We will then perhaps have a better chance of predicting the physics that's observed experimentally.

In these four-dimensional theories, is there a sense in which the other six dimensions of the ten-dimensional theory are still there but in a different form?

The situation is much more profound than that. Really what is happening is that in any string theory there aren't four or ten dimensions. That's only an approximation. In the deeper formulation of the theory the whole notion of what we mean by a dimension in spacetime will have to be altered. Our normal concept of spacetime is that it is a smooth collection of points. Any position in space or time is specified by a point. One ought to formulate string theory in a much larger space – something like the space of all possible positions of a string. In fact, this is an *infinitely* larger space, so that when we are talking about a theory which works in ten dimensions or four dimensions, we are actually talking about an approximation to an infinitely larger structure. In terms of this infinitely larger structure, the distinction between a formulation in four dimensions or in ten dimensions is much less. The reason we use the language of ten dimensions or four dimensions is because we have so far been forced to talk about string theories in an approximate way, and it's only in this approximation that the whole notion of a small finite number of spacetime dimensions makes sense.

Has the idea of higher dimensions getting curled up, and the interest in the way that they get curled up, gone away now?

By no means. It's certainly true that certain aspects of that are less emphasized than others, but it has by no means gone away. In fact, in a certain sense, it has all been subsumed within this much larger structure. Talking about four or ten dimensions at all is itself only an approximation to this much larger stringy space which really has an infinite number of dimensions.

So one has still got to find a way of rolling up higher dimensions with this stringy space?

Well, in the stringy language, whether we have six extra dimensions or not, is less relevant given the fact that we now have infinite numbers of dimensions. The framework of talking about curled up dimensions has now broadened into a much bigger framework which involves trying to understand what it means to have stringy space-time and to what extent the physics we see emerges as an approximation to this much richer structure.

Do you think we are ever going to be able to envisage what we mean by stringy space?

Well once the correct fundamental formulation of the theory is really understood, it will probably be something that is startlingly simple. That's the usual situation in physics. When one first discovers an intriguing new structure, things seem very complicated, but when it's really understood, things get clarified and simple. Of course, I don't yet know what this new formulation might look like but certainly the hope is that it will be something which is simple. Whether or not one could envisage it in very concrete common-sense terms, or whether it would only look simple to one who is well educated in sophisticated mathematics is a question to which I also don't know the answer.

Turning now to the scientific status of string theory, Feynman is quite negative about the superstring theory because he says it's failing to make contact with experimental data such as the masses of various elementary particles and the strengths of the coupling constants. What do you say to that?

I would not have thought that this was the kind of approach to physics that Feynman would favour. I think it's fair to say that at the present time the theory is some way off making very detailed predictions about specific measurements of elementary particles. People are working very hard on trying to understand the predictions and I have no doubt that more will be understood eventually.

I have said before that the way in which the theory has been understood so far has been in terms of successive approximations, but there are certain questions which quite clearly cannot be answered until we can go beyond this approximation scheme. The question of the masses of the particles which we see is one example of this. At the present level of approximation all particle masses are zero: all particles are massless in this approximation. Now that, in fact, is a good approximation when you realize that the scale on which we are measuring these masses is the so-called Planck scale. This is 10^{19} times the mass of a proton, so that the mass of anything that we have seen in the laboratory is very small on that scale. It's therefore a good approximation to say that masses are zero.

Of course, the particles we actually see around us are not massless, they weigh something, and we ought to be able to predict their masses. That sort of prediction, the fact that these particles are not massless, and the values of their masses, is the kind of prediction that is very difficult to make in string theory in its present formulation.

Other very interesting questions also cannot be answered until we understand the theory better. Questions such as how we describe black holes in this theory. This is a theory containing general relativity, therefore it must contain black holes. How does one describe them in terms of strings? Other questions relate to the cosmology of the early universe. There was an era in the early universe, when it was extremely hot, where string physics was very important. In order to understand what string theory has to say about the evolution of the early universe one would have to understand it, again, in a way which goes beyond the approximations that are so far being used. So for many interesting questions, the theory isn't yet understood well enough to provide the answers.

My own view is that these are still early days and the success of the theory in any case ought not to be judged on whether or not it can predict details of things that we have already measured. Surely if such a completely new kind of theory is correct, then it represents a change in the structure of physical theories which is big enough to have some sort of implications for things which we really haven't thought about measuring yet. There ought to be new kinds of predictions which should be quite startling.

Do you believe there are going to be?

I certainly don't think that we know yet. We are not yet able to deduce all the predictions of the theory, but already there are certain ideas, which I must admit are not really compelling as measurable and firm predictions of the theory, but are nevertheless striking in their nature. For example, one prediction that might come out of these theories is that there should be a whole new type of matter in the universe. This has been called shadow matter – matter we would not be able to see directly except for its gravitational effects on us, though particles of shadow matter might exert strong forces on each other.

This shadow matter could exist all around us, is that right?

It could. I'm not saying that I believe the theory predicts it, but it's certainly a possible prediction of the theory.

Are you suggesting that there are, in some sense, two copies of the universe, the one we inhabit and another shadow universe which we don't see except possibly through gravitational effects?

Well, let me cautiously say that the theory *might* predict that. However, whether or not this stuff actually exists around us depends on the details of the history of the universe which are very difficult to calculate in any case.

Presumably if a shadow star or planet passed through our solar system we would notice.

Yes.

It's a bit tricky, though, to test a theory by looking solely for those gravitational effects.

That's right. Even if this matter existed, that is not an example of a prediction that would be very easy to test.

Can you give me any other examples of one of the predictions of superstring theory that might be worth testing experimentally?

At the moment there are no firm predictions that we know about. However, a possible prediction has to do with the fact that in string theory there might be extra dimensions with curious topology. There may be extra dimensions with holes in them, like a doughnut. Then an object like a string can get trapped by winding around such a hole. Such trapped strings would have strange properties. For example, they could give rise to what we would see as new kinds of particles, which would be very heavy and which would have curious electric charges – fractional, noninteger electric charges. These particles would be too heavy to make in the laboratory but could have been created in the big bang when the universe was very hot.

I should emphasize that this is one rather way-out suggestion of the sorts of effects that string theory might have which are very different from the conventional particle theories. So although this sort of prediction shouldn't be taken too seriously at the moment, it indicates that there are ways in which string theory is different from conventional theories.

These are early days yet and we will hopefully find other effects which are equally distinctive.

Sheldon Glashow has also been very critical about string theories. He says that they could undermine the motivation for doing experiments by conveying the impression that the theorists have already wrapped the subject up. What's your response to that?

I'm sympathetic to the view that these theories are at present very remote from being able to explain directly what is measured experimentally in accelerator laboratories. Given the fact that they are so very different from previous kinds of theories, then they ought to predict some entirely new sort of phenomenon that we haven't even thought of measuring. It was only *after* Einstein had formulated general relativity that he understood which phenomena that could be measured, would test the theory. The precession of the perihlion of the planet Mercury was already known, but it wasn't until Einstein came up with general relativity that it was realized that this peculiar anomaly in experimental measurements was of fundamental importance. So what we need in superstring theory is the analogue of the planet Mercury. Some distinctive piece of experimental evidence that might already be known but hasn't struck anyone as being important because no-one realizes that it's of relevance to testing a fundamental theory.

Glashow, as I understand it, argues that string theorists are approaching physics in fundamentally the wrong way, that they are adopting what's sometimes known as a top-down approach. They are starting with a general formulation and then trying to work from there towards a description of the real world. Glashow prefers to start with the findings of experimental physics and somehow gradually build up a theory, perhaps work towards a general theory, but starting from experimental physics. Do you think that there is a dichotomy here that you should be worried about?

Well, I think there's room for both approaches. Historically theoretical physics has advanced by both methods and you can call on historical precedent for either approach. Certainly I would agree that the motivating force behind the research being done in superstring theory has been, and still is, the elegant theoretical structure, and the hope of resolving what I think is *the* fundamental theoretical paradox in physics in this century, which is the conflict between quantum mechanics and general relativity. This has certainly been my motivation and other people's motivation.

I think it is important that there are people around who pursue the so-called bottom-up approach as well. The two groups of people can easily coexist and presumably help each other.

Looking back at the early days of string theory when not so many people were interested in it, was there a time when you felt as though you had been positively shunned by other physicists because you had been dabbling in this area of physics?

No. I don't think we were shunned. I think that we were, to a large extent, ignored, partly because string theory is so very different both conceptionally and technically from the kinds of theories that were then fashionable. In the early 1980s it would have required so much effort for people who were not working on string theory to learn the techniques, to decide for themselves whether they believed in the theory, that people were just not willing to make the effort, with a few exceptions. In a sense life was very nice in those days because particle physics is generally a very competitive subject and it was just nice to be working on something that we could take at our own pace without feeling pressurized.

String theory, of course, had had its earlier phase in the early 1970s and by the time it was dying out, around about the mid-seventies, it was definitely not a subject that one ought to be doing from the point of view of one's reputation. Probably this applied more in the United States than in Britain, but certainly the mainstream of particle physics – the sort of physics that the important people were doing – was very much *not* string theory and it would have been difficult at that time to get a job if one was working on string theory. I think this was reflected by the fact that there just weren't any other people at that time working on the subject.

How did you actually begin your collaboration with John Schwarz?

We had known each other some time earlier but never worked together until the summer of 1979 when we both happened to be visiting CERN. CERN is a wonderful place for people to get together and exchange ideas, and we were just talking about supersymmetry and strings which had been things which we were both interested in, and this developed into our collaboration.

Looking now to the future, Ed Witten has said that superstring theory is a twenty-first century theory that has dropped by accident into the twentieth century, and that he thinks it's going to dominate the next fifty years of physics. Do you see the subject in the same way?

Well, I'm pretty sure that developments from the superstring theory will become the predominant trade of theoretical particle physicists for a long time. I would actually prefer to say it another way. I cannot imagine how anyone who has previously worked on general relativity, for example, and who is now working on string theory, would ever go back to working on general relativity without strings. That seems inconceivable.

Will superstrings turn out to be the Theory of Everything?

Let me just say that it's because we really understand so little of the deep structure of the theory, that I object to this terminology that's often used, that this is the Theory of Everything. We just don't know what the theory predicts yet and we also don't know yet what questions ought to be asked. My feeling is that by understanding the theory in a much deeper way, all sorts of issues and questions will be raised to which the theory at that point will probably have no answer. So I think saying it's a Theory of Everything, is merely saying it seems as though it might answer the questions that we now think are important in particle physics.

It is a theory which at least pretends to address the issue of how all of the particles and all of the forces are related together, is that correct?

Yes. It addresses that question and, in fact, it's obviously giving some very interesting indications as to what the answer is.

So it does amalgamate the forces, the matter that the world is made of, and the space and time which encompasses it. That sounds like everything to me!

But we don't yet know how to formulate the theory in a way which does unify spacetime with string-like particles. We don't know what the theory has to say about physics beyond the Planck scale that plays such an important role in our present ideas.

So there may even be a deeper level?

There may be a whole new set of issues, a new set of questions which the theory can't answer, and I don't think we even know what the questions are until we understand the theory in a more logical way. It's plausible, for example, that this will entail a radical change in our ideas about quantum theory. That would be very exciting.

What about going the other way, not to a deeper level, but looking at larger and larger scales where one has systems of ever greater complexity? One might also object to calling this a Theory of Everything because it wouldn't, for example, explain the origin of life.

That's right. There are all sorts of issues of complexity that understanding of physics on the microscopic level probably is of little relevance to.

But would you agree that, if successful, superstring theory would represent the culmination of a two and a half millennium search for the ultimate building blocks of reality, the triumph of the reductionist programme?

I am not, myself, of the opinion that there are 'ultimate building blocks'. I just cannot believe that in two billion years time nobody will have come up with a better theory. I certainly think that it's a good theory for the time being and it will last for many years. The fact that string theories relate to so many branches of mathematics indicates that they contain deep truths.

So strings are here to stay?

For a long time.

5

David Gross

David Gross is Eugene Higgins Professor of Physics at Princeton University. He is a leading elementary particle theorist, and has made important contributions to the subject of quantum chromodynamics. As one of the so-called Princeton string quartet, he is one of the originators of the so-called heterotic string model.

One of the odd features about the superstring theory is that it has to be formulated in more than four spacetime dimensions, which means that there are dimensions of space that we somehow don't see. Could you say a little bit about how that works?

The idea that there might be more than three spatial dimensions is a very old one and not peculiar to string theories – although string theories differ from other theories in that they must be formulated in more than three spatial dimensions. At first that was regarded as very bad, but now we realize that it really is an experimental issue how many dimensions of space there are. If the extra dimensions are curled up into little circles (or more complicated surfaces), and are small enough, we would never know that they exist on the basis of casual inspection.

Let me make sure I have got this absolutely right. Are you saying that what we normally think of as a point in ordinary three-dimensional space is, in fact, some little bundle of higher dimensions?

Right. A straw seen from far away looks like a line, but when we get up close and have good eyes or good magnifying glasses, we can see that it has an extra circular dimension. In the same way every point might have extra dimensions, in some as yet unexplored directions. In string theory we need six of them, and if the theory is to be in agreement with the fact that we haven't noticed them yet, they must be curled up and very small. The fact that they might turn out to be small is reasonable since the theory has a natural length scale which is very, very small (10^{-33} centimetres). It is certainly conceivable that in a theory like this the extra dimensions will automatically curl up and only provide us with three spatial directions which are big and open.

Supposing we had the apparatus to be able to look at this very fine level of detail and see these extra dimensions, what would they look like?

Well, how do we actually look? The way that we look is that we build enormous accelerators, and those accelerators probe physics at very short distances.

This is quite hypothetical?

Yes, a hypothetical accelerator, which would be of the order of 10^{16} times more energetic than present day accelerators, and cost 10^{20} times more money than we could possibly raise, would be able to probe these extra dimensions, but in no way would we be seeing them as we would, say, under a microscope. If we could imagine doing that, they would look like right, left and up – they would just be right, left and up in six other directions, except that in those other directions one would go around and come back to the same point – they would be circular and closed in those directions.

Is it possible to tell from calculation what the shape of this additional six-dimensional space would be?

Ever since Einstein, the question of the geometry of space and time has been a dynamical issue. It should be determined by the physics. In principle, therefore, one would take a string theory and solve the string equations. Then the solution of that theory, being a theory of the structure of spacetime, would determine what geometry space and time possessed.

Now in practice, what one has done so far within the context of the heterotic string theory is to explore the possible classical (that is, nonquantum) solutions of the theory. One takes the theory and somewhat indirectly deduces the possible solutions of the equations of motion of the theory. And remarkably enough for the heterotic string, we have found a whole class of solutions, in fact millions and millions of possible solutions. Some of these describe a world which, geometrically, is like our own. There are three spatial dimensions, one of time, and six small compactified dimensions which curl up into rather unusual exotic mathematical manifolds or surfaces, with properties which mathematicians delight in, and physicists have had to learn. So to that extent we have learned that the heterotic string theory has consistent solutions which geometrically resemble our world. It also has consistent solutions which do not look like our world, which have more than three open spatial dimensions, and we don't know yet what principles of physics pick the four-dimensional solutions from the ten-dimensional, the six-dimensional or the eight-dimensional.

Are there many solutions which have three space dimensions?

Yes, there are millions and millions of solutions that have three spatial dimensions. There is an enormous abundance of possible classical solutions. Not only are these solutions okay classically, they seem to be okay quantum mechanically. When one looks at quantum mechanical corrections to them which could have very well led to nonsense or instabilities, one finds that that doesn't happen, to all orders of perturbation (where one assumes that the classical solution is basically right and all you have to do is make slight quantum corrections).

This abundance of riches was originally very pleasing because it provided evidence that a theory like the heterotic string theory could look very much like the real world. These solutions, in addition to having four spacetime dimensions, had many other properties that resemble our world – the right kinds of particles such as quarks and leptons, and the right kinds of interactions. These emerge from the theory naturally, or at least could emerge from the theory naturally. That was a source of great excitement two years ago.

It is, however, slightly embarrassing that we have so many solutions but no good way of choosing among them. It is even more embarrassing that these solutions have, in addition to many desired properties, a few potentially disastrous properties. These include symmetries of the theory that do not appear in the real world, and so must be broken somehow. Then there are massless particles that have never been observed, and, in fact, are ruled out by experiment. So something is wrong with all of these solutions that we have so far. We would like to believe that dynamical effects that are not revealed in this perturbative approach will cure these problems, and make it possible to pick one, and only one, solution of the theory from all of these as yet equally good solutions.

Let me see if I've got this right. The theory is, at the moment, being tackled in a certain approximation scheme – perturbatively – by considering it to be a series of small corrections, and that all of these approximate solutions appear to be unsatisfactory in some way. Not only are there far too many of them, there is also the problem that none has completely satisfactory features. But you are suggesting that if one could tackle the mathematics to produce an exact *solution then this ambiguity would go away?*

Right. That is the case in many other theories that we know, for example in quantum chromodynamics, which is a theory of quarks and gluons describing the nuclear force and the structure of nuclei. The properties of hadrons (nuclear particles) can only be arrived at by a very complicated nonperturbative mechanism. A perturbative approach to the theory, similar to the one we've used in string theory, yields total nonsense.

In string theory, to date, we only know how to use a perturbative treatment. We don't yet have an adequate understanding of the theory, or even an adequate formulation of the theory, that would allow us to tackle nonperturbative questions. But, it is extremely unlikely, for a variety of reasons, that perturbation theory is sufficient.

What are the reasons for that?

Well, firstly, if the theory is correct, it had better not be sufficient because a perturbative treatment disagrees with experiment!

Secondly, string theory *contains* many theories that we are acquainted with, like quantum chromodynamics, for which perturbation theory is known to be insufficient.

Thirdly, this is a theory with no arbitrary parameters, no adjustable constants. If you find a solution to the theory there is nothing you can fiddle with. It is unique. Everything is calculable. It is very unlikely in such a theory that you could have a perturbation series – what quantity would you be expanding in? Normally you can expand in something when you have a small constant that you can adjust, but here there is no small constant. Every constant in the theory is calculable.

Fourthly, if you are to come up with a theory of this type which seems to contain all of physics, it will have to deal with a very fundamental problem in physics, namely the cosmological constant problem.

Tell us about that.

That is the issue of the background energy of the universe. In ordinary theories of matter where you neglect gravity, the absolute scale of energy is immaterial. You don't care. All you care about are energy differences; there is no way of measuring the absolute scale of energy. However, gravity is a force that couples to energy. Normally we say that gravity couples to mass, but mass, as we know from Einstein, is energy. Gravity couples to energy and in a sense 'knows' how much energy a given object has – and that also applies to the universe as a whole. The universe itself has some energy density.

Even when space is empty?

Even in empty space. You can measure the energy density of empty space because the more energy density you have, the more the universe curls up due to the attractive force of gravity. So by measuring the global structure of the universe, you can measure the background energy density of the universe. And it *has* been measured – not exactly, but bounds have been placed on it because it seems to be very close to zero. In fact it seems to be the best experimental determination of a zero quantity we have ever come up with! It's zero to an accuracy of one part in 10^{120}; that's in units of the Planck mass – the natural mass/energy scale of gravity. What this means is that if you were to sit down and work on any of the current physical theories that include gravity, and if someone were to ask you, in the absence of any observation, your guess as to the background energy density of the universe, it would be 10^{120} times bigger than the bound on what is actually observed. What is observed is so small, in fact, that everybody believes it must be zero. But there is no reason why it has to be zero! As I said, naturally it should be much bigger. Not only that. Even if you adjust it to be zero in the theory, if you simply arrange for the theory to have zero energy density, by hand (something physicists don't like to do if they have to adjust something to 120 decimal places), even if you do that, and then you find you've forgotten some small quantum mechanical effect, then this is bound, as far as we know, to generate a sizeable cosmological constant again. The smallness of the cosmological constant has been a mystery since Einstein first introduced it. Since then it has been found necessary to set its value equal to zero, zero, zero and nobody has understood why.

Now that's alright, as long as you don't claim to have a Theory of Everything. But if you purport to have a Theory of Everything it has to solve this problem as well, because the Theory of Everything will either produce a cosmological constant, or not. If it does not, and yet still reproduces what we see around us, in the real world, that will require some physical mechanism that we do not at this time understand, and certainly not one that is likely to be treatable by perturbation theory.

So far in string theory the cosmological constant has remained at zero. That is, there are solutions of the heterotic string theory which produce four observable dimensions, and those four dimensions look like our world; that's to say there is no cosmological constant. If there had been we wouldn't have had four big dimensions that we could walk around in. In fact those three spatial dimensions would be curled up into a little sphere smaller than an atom. That doesn't happen. The reason it doesn't happen is understood to be related to supersymmetry – to the fact that these superstrings are supersymmetric – and it's that symmetry which prevents the cosmological constant from developing. We have no idea how that symmetry can be broken (it must be broken because it is not evident in the world) and yet not produce a cosmological constant. Any mechanism that anyone has ever thought of for breaking supersymmetry, also produces a cosmological constant.

So there is something very strange going on in the physical world, some new principle or new way of breaking supersymmetry that is going to somehow solve this problem and if it is to be string theory that does that, it has to be by a dynamical mechanism that's quite different from what occurs in perturbation theory.

Don't you think then that the solution to the cosmological problem will be built-in in a fundamental way into the string theory?

It could very well be built-in in a fundamental way into the string theory. There is no evidence for that because the conclusive evidence would be a unique solution. However, in the theory so far the cosmological constant is zero as far as we know. At the same time, supersymmetry is unbroken. We have these two things going on together and they seem related. One is good and one is bad. Now we would like to hope that the theory manages to break supersymmetry and not produce a cosmological constant of any magnitude. There is no evidence that that will take place, only the hope that the theory describes the real world. But if it does, then one is going to have to discover some very fascinating new dynamical process or mechanism that is not adequately described by the perturbative treatments that have been given to date.

What, then, is the way forward, given that the perturbation treatment is a fairly direct one, and in mathematical terms much easier than an exact treatment? Are you simply going to have to learn some new mathematics?

Well, that is the direction that most people are taking nowadays. There is a lot of physical stimulus as to what one wants out of the theory. So far as we have seen in these perturbative solutions the theory has most of the ingredients we need to account for what we observe at low energies. What we are missing are a few extremely difficult answers to some of these fundamental questions.

So where do we go from here? In normal science, the way particle physics was ten years ago, we would wait until our experimentalist friends provided us with a clue. That's the way we always made progress in the past. Well, we don't have that luxury any more.

There is just not enough money to build the accelerators large enough to do this?

There is not enough money in the treasuries of all the countries in the world put together. It's truly astronomical. Not only is there not enough money, there is also no conceivable way known today of building such accelerators. We hope very much that we will be able to build accelerators ten times greater than the ones we have today to explore some of the interesting physics in the next energy domain, but the push up to the Planck mass is impossible in the foreseeable future, if ever. So there will be no direct clue coming from the relevant energy scale. We have to look for indirect clues from cosmology or low energy physics, and more and more we are forced to look for mathematical suggestions to explore the ramifications of the theory and look for new mathematical structures. This is a chancy and dangerous procedure for theoretical physics, but if there is no other way, we'll have to follow this course.

Of course a sceptic might say that there have been previous occasions in history when physicists have thought they had everything wrapped up in a single unified theory, and this has always turned out to be wrong. Is there a danger that you are just chasing after a mirage here?

There's always a danger that we are chasing after a mirage even when we are working in conjunction with an experiment. That danger always exists and so you have to try very hard to test your ideas all the time so that you won't spend a lot of time going down blind alleys. There is a feeling among some of us that this time it is a little different, but that could easily be wrong. The structure of these theories is incredibly rich and in many ways contains the knowledge we have already, or at least appears capable of containing low energy physics as we know it, in a way that hasn't really been true of grandiose theories before. But it could be illusory and it could be that there is something even stranger out there than superstrings and ten dimensions which is necessary to account for everything. There is no way of telling without giving it a try, and that try is going on, and will go on for years until there is either evidence that it is fundamentally wrong, or someone comes up with better ideas. Actually better ideas are usually more important than contradictory evidence, because physicists have to work on something! In this case, if there is no better competing idea around they will work on strings.

The string theory does seem to provide a peculiar attraction for theoretical physicists. I don't think in my experience I've seen such compulsion for a theory. Is there something intrinsically satisfying, or intrinsically promising about using a string as the structural basis for a Theory of Everything?

There are two reasons why it has become such a popular field in the last two years. The most important one is that there are no other good ideas around. That's what gets most people into it. When people started to get interested in string theory they didn't know anything about it. In fact, the first reaction of most people is that the theory is extremely ugly and unpleasant, at least that was the case a few years ago when the understanding of string theory was much less developed. It was difficult for people to learn about it and to be turned on. So I think the real reason why people have got attracted by it is because there is no other game in town. All other approaches of constructing grand unified theories, which were much more conservative to begin with, and only gradually became more and more radical, have failed, and this game hasn't failed yet. Furthermore, it was realized from the beginning that it could potentially do a lot more than all the other approaches could possibly achieve.

The second reason for the attraction of string theory, is that as you study it, and as the theory develops, more and more people are convinced by its beauty. It is a very beautiful theory even though it's only rather primitively understood, so it's likely to be even more beautiful once we understand it at a deeper level in the future. So at this point in time no other new ideas have come along and people are more and more impressed with the depth and structure of the theory.

I'm speaking to you in Princeton, the former home of Albert Einstein. If he were alive today what do you think he would make of the superstring theory?

Well, one always wonders what Einstein would think of various things. I have asked myself that question many times, what Einstein would have thought of this, what Einstein would have thought of that. Of course, you would have to get Einstein to overlook the fact that this is a quantum mechanical theory, you would have to explain to him about supersymmetry, which is a sort of marvellous extension of his notions of space and time. I think he would have liked supersymmetry. I can't imagine him not liking supersymmetry. It doesn't require quantum mechanics. In fact, in many ways this extension of space–time symmetries is a partial realization of Einstein's goals. Einstein had two goals. One was probably misplaced, namely that quantum mechanics would emerge dynamically from a highly constrained classical theory which would provide, because of the tight constraints of the equations, quantization conditions. Nobody today believes in that. People believe that quantum mechanics is for real and is here to stay.

But Einstein also believed that geometry would determine dynamics. He used to make the following remark about his famous field equations. The equations of general relativity are such that on the left-hand side there is the curvature of spacetime, and that is equal on the right-hand side to the energy and momentum of matter that serves as a source that curves space and time. Einstein used to say that he liked the left-hand side of the equation – that was beautiful, that was geometry, that was the curvature of space. But he didn't like the right-hand side, which referred to this 'matter' that you had to put in in an arbitrary way. So he used to say the left-hand side of his equation is beautiful and the right-hand side is ugly. Much of what he was doing in the latter part of his career was trying to move the right-hand side to the left-hand side and understand matter as a geometrical structure. To build matter itself from geometry – that in a sense is what string theory does. It can be thought of that way, especially in a theory like the heterotic string which is inherently a theory of gravity in which the particles of matter as well as the other forces of nature emerge in the same way that gravity emerges from geometry. Einstein would have been pleased with this, at least with the goal, if not the realization.

He would have liked the fact that there is an underlying principle unifying all of physics, presumably.

He would have liked the fact that there is an underlying geometrical principle – which, unfortunately, we don't really yet understand.

6

John Ellis

John Ellis is a theoretical physicist at the Centre Européen pour la Recherche Nucléaire (CERN) near Geneva, and has played a prominent role in the formulation of supersymmetric and gauge field theories aimed at unifying the forces of nature. He is well known for his attempts to link new ideas from particle physics, and most recently superstrings, to observational cosmology.

Could I start by asking you to give a brief summary of what you believe the superstring programme aims to achieve?

I think that the superstring is the first serious candidate that we have for a unified theory of all the fundamental interactions in nature, going all the way from gravity, which is responsible for keeping the planets in orbit around the Sun, through electromagnetism which keeps electrons in orbit around nuclei, through the strong interactions or the nuclear force which obviously is very important in holding nuclei together, and also the weak interactions which are responsible for many forms of radioactive decay. Up to now some of these interactions have been unified in a partial way but there has been no real solid claim to be able to unify all of them in a single mathematical picture.

What is the essence of the theory?

According to the superstring idea, all the particles which we previously thought of as elementary, that is, as little points without any structure in them, turn out in fact not to be points at all but basically little loops of string which move through space, oscillating about.

What exactly are these strings? How should we envisage them?

Let's consider the old picture of an elementary particle. In that case you just have a point, and then as the particle moves through space you can imagine it describing a line, called a 'world-line'. Now, in the superstring theory what happens is that at any moment in time this particle is actually a little loop, you can imagine it being a lasso, or something like that. Then as time develops, this little lasso moves through space and so it describes something which is rather like a tube going through space and is called a 'world-sheet'. That is the trajectory of a particle according to the superstring idea.

So we must think of a particle as really being an extended object which can also have some type of internal motion as well. Is that right?

That's right. When we consider atoms, we know that they are made out of constituents; there are the electrons going around the nucleus in the middle, and then of course the nucleus itself is made out of constituents called protons and neutrons. Protons and neutrons are in turn made out of constituents called quarks. According to the superstring theory, the quarks are also extended objects, but they are not really made out of any more fundamental constituents. I mean, there aren't little subquarks sitting inside the quarks. They are made of this piece of string but it doesn't have any internal structure – it has a size, the typical size of this loop of string we believe would be something like 10^{-33} centimetres, that's one thousandth of one billionth of one billionth of the size of a nucleus, roughly speaking.

If it's the case that all of the different particles are made out of these little loops, how is it that we get differences between the particles? Why are there so many different sorts of what we previously thought of as fundamental objects?

I think that it's good to think in terms of the classical strings that we know and love, like, for example, violin strings. Now you know when you pluck a violin string it can oscillate at different frequencies – it has different harmonics. The superstring is something like that. The different types of elementary particle, we believe, will correspond to different ways for this loop to oscillate, rather like different notes that you can play on the same violin string. In principle there is, in fact, an infinite number of different ways for this superstring to oscillate. The elementary particles that we actually see nowadays, the things that we are made of, will just correspond to the lowest harmonics, much like the lowest note which you could play on a given string.

Are you saying that the difference, say, between an up quark and a down quark is more or less entirely due to the different pattern of motion going around this little loop?

That's right. This superstring, in addition to oscillating in space, rather like a traditional violin string, also has certain internal degrees of freedom which you can't really visualize in terms of simple oscillations in space, and the actual difference between, say, an up quark and a down quark would presumably be some sort of a combination of these internal properties and these oscillations in space.

Is it conceivable that if we had sufficiently powerful instruments, we could actually directly probe these little loops – actually manifest them rather than simply believe that they're in there somewhere?

In principle, yes. But in practice I think it would be very, very difficult. In order to really see this loopy structure inside particles, you would have to do experiments which probed energies of 10^{19} GeV, which is some ten million billion times more than the energies which we have achieved up to now in our particle accelerators. I'm afraid that building an accelerator to do something like that would be unimaginably expensive and probably we don't have the technology to do it anyway.

I'm sure you're right, but if we could achieve those energies, would it be possible to actually snip through these loops and open them up and have open strings rather than closed loops?

My belief would be that that's probably not possible, though that's a matter of opinion. Some people think that the strings could actually be snipped open, as you say, and there may be open strings as well as closed strings. I, myself, tend to prefer the theory according to which you only have closed strings.

But it could very well be that as you heat matter up to incredibly high temperatures, the whole string could just unravel and dissolve. But that for the moment is just a speculation which we are not in a position to justify.

One other point concerning these little loops, in the case of charged particles, do we imagine that the loops carry electric charge, and that this charge is distributed uniformly around them?

Well that comes back to the point that I was trying to make earlier on. You shouldn't think of elementary particles like, say, an electron which has a charge, as containing more elementary constituents which carry the individual subcharges which add up to make the total charge of the electron. In fact, what we call electric charge would be some sort of collective property of the string as a whole and if the string oscillated in different ways then it would seem to have a different electric charge.

In other words, the electric charge might be seen as a quality of the motion of the string rather than something which is just added on to a particle or fundamental object.

Yes, I think that would be a good way of thinking about it.

People often ask what electric charge is, and usually you can't say anything other than it just is a fundamental property, but you seem to be saying that we could explain electric charge in terms of some sort of dynamics.

Let's think back to what we actually mean by electric charge. What we mean is that there is a field called the electromagnetic field which is the thing coupling to electric charge, and it is the electromagnetic field which is responsible for holding electrons around nuclei, or is responsible for radio waves, for example.

The electromagnetic fields are, in fact, themselves associated with particles called photons. These photons, again, are a different mode of oscillation of the string, in just the same way that the electron is some mode of oscillation of the string. So what we think of as electric charge is really a coupling together of different pieces of string which are oscillating in slightly different ways, and the photon is neither more elementary nor less elementary than the electron.

One of the unusual features of the superstring theory is that these little loops live not in the usual three dimensions of ordinary space, but in a ten-dimensional spacetime. Why is this necessary?

It turns out that to formulate the string theory so as to be consistent at the level of evaluating quantum corrections, and if the string has no other internal degrees of freedom, we have to formulate the theory in a special number of dimensions. If the theory only contains what we call bosons, which are particles like the photon which have integer spin, then that critical number of dimensions for formulating the theory turns out to be twenty-six. I'm afraid I can't give you a very simple explanation of that, it's just how the mathematics turns out.

Now, if we complicate the theory a bit, and put in fermions as well (fermions are particles of half-integer spin like, for example, the electron), then the critical dimension turns out to be ten. Of course, that's still rather a long way from the three dimensions of space plus one dimension of time which we seem to be living in.

How then can we make this theory consistent with the fact that we actually only perceive three dimensions of space and one of time if there really are ten dimensions?

One approach is just to say okay, there are six surplus dimensions of spacetime, or maybe twenty-two surplus dimensions of spacetime, which we have to get rid of. Then you curl them up rather like crumpling up a piece of paper. You can imagine that a normal piece of space is analogous to a piece of paper, set out flat, and then if you roll it up so as to make a little tube – like you might do with a newspaper if you wanted to put it in a dog's mouth, for example – that would correspond to taking the original two-dimensional surface of paper and in some sense compressing it to one dimension, the one dimension corresponding to the length of the tube.

So you could imagine doing something like that to the extra dimensions of spacetime. You curl or roll up the surplus six or twenty-two dimensions, leaving the four dimensions which would correspond to the length of that newspaper tube.

There must be many different ways of doing that?

There certainly are. What people did when they first tried this approach was that they wrote down a set of conditions which seemed to be necessary for this rolling up mechanism to be consistent, and although those conditions were quite stringent, nevertheless, there seemed to be something like ten thousand different possibilities for rolling up the newspaper. At the moment some physicists are exhausting themselves checking through all these ten thousand different ways of rolling it up to see whether any of them look like the real world.

You mean they would all lead to different types of physics at the sort of energies which we can observe?

That's right. For example one of them might give two photons instead of just one photon. Or one of them might even give *three* photons instead of just one photon. Or some of them might, instead of giving three electrons which is what we see in the real world (when I say three electrons I include the muon, which is very similar to it but just weighs a bit more, and the tauon which is again similar but weighs even more still) give a fourth electron which would weigh even more than the tauon. But there are various reasons for thinking that this is unlikely to be the case in the real world, so people try to formulate theories which only have three electron-like particles, and which only have one photon.

So the number and nature of the particles and forces is related to the way in which these higher dimensions are curled up – the different possible topologies?

That's right. In fact you may be able to relate the number of electron-like particles to the number of holes which you have when you roll up these extra dimensions. If we think in terms of the newspaper analogy, for example, you might say that the rolled-up newspaper has one hole. If you look along the tube of the newspaper there's just one hole going through the middle of it. On the other hand you could imagine (at least in some sort of abstract sense) rolling up the newspaper in such a way that there was more than one hole in it. Now according to the mathematics of these theories, it would be the number of such holes which would determine for you the number of electron-like particles.

It seems that we begin to explain the world in terms of topology – the shape or the connectedness of this higher dimensional space – rather than in the traditional way of supposing that things were just made that way.

That's right. I said earlier that in the superstring theory you don't regard what now seem to be elementary particles as being composed of still smaller objects. One way people used to imagine that you might be able to get different sorts of electron-type particles was to take these subconstituents and combine them together in different ways to get different electrons.

But that's not the way it works in the superstring. In the superstring these different types of electron, as I said before, will correspond to the different holes which you could obtain when you were rolling up your imaginary newspaper.

Coming back to the different topologies, you seem to indicate that at the moment it is completely unclear as to which particular topology would correspond to the real world, and that there are many, many contenders. This seems like a real weakness of the theory, because obviously it's not selecting a unique possibility. Is this just a matter of ignorance? In other words, with further investigation is it possible that we'll discover that there is a unique topology which gives us the real world, or will there always be some ambiguity?

I don't think we know the answer to that question yet. One possibility is that when we understand the theory better, we will figure out that there is only one consistent way of rolling up the newspaper, and that must be the way that the universe is.

But another possibility is that actually there are many different consistent ways of rolling up the newspaper, and that different pieces of the universe have chosen to roll up the newspaper in different ways. That could mean that, for example, there will be some region of the universe out there somewhere or other which maybe does have two photons or maybe it does have four electrons. At the moment I just don't think that we can decide between these two alternatives.

And perhaps the reason that we see the particles that we do has something to do with the fact that we couldn't be here – that is, life couldn't form – unless the universe were more or less as it is.

I think that if, for example, the number of electron-type particles were different or the number of photon-type particles were different, it would probably still be possible to build something which would be interesting as far as the universe is concerned. It wouldn't look exactly like our own universe, and maybe the physicists sitting around discussing the structure of the universe would look rather different from the way we look, but I still think that probably in many of the cases physicists could still exist.

This rolling up business – is that something which we should be able to understand dynamically? Are there forces in some sense causing the higher dimensions to curl up, or is it just an abstract mathematical thing which is intended?

Well, there are forces in some sense. The string itself, for example, has a tension. Let's think back to the violin string that we were talking about earlier on. You can adjust the tuning of a violin string by varying the tension in that string. Now similarly to the violin string, the superstring also has a certain intrinsic tension. In that case the intrinsic tension is provided by the underlying structure of the theory, but it does have a rather similar property. So this tension would be some sort of intrinsic force which the string itself has.

Is this force in some way contributing to the curling up of the extra dimensions?

This string tension does play an important role. In terms of the various different topological configurations of our newspaper, there is obviously some barrier between the different ways of rolling up the newspaper. There is something which prevents the newspaper from spontaneously unrolling, but we don't know what that is in the case of the superstring. Also, we do not know on purely theoretical or *a priori* grounds, how large the radius of this 'tube of newspaper' should be. Maybe it's 10^{-33} centimetres as I suggested earlier on, or it could be 10^{-34} or maybe 10^{-32}. As things stand we don't have any way of calculating the absolute size of this rolled up piece of 'newspaper'. Hopefully we are going to find some way of calculating it one of these days. It may be that it has something to do with higher order effects in the theory such as quantum corrections, like the Casimir effect, for example which creates a force between electrically conducting plates. It may be that something similar is operating in the superstring, but that hasn't been demonstrated yet.

It seems, then, that the dynamics of this spontaneous compactification – this rolling up of the higher dimensions – is a subject which is not at all understood.

I think that's absolutely correct. It may well be that this whole idea of compactification is going to be thrown out next week. Nowadays some physicists are toying with the idea of not formulating the string theory in twenty-six or ten dimensions but actually formulating the thing directly in four dimensions and making no reference to the possible existence of extra compactified dimensions.

How would that be possible?

Roughly speaking, what you do is to trade in these old-fashioned spacetime degrees of freedom, these extra surplus dimensions, and you replace them by coordinates in a purely internal space, rather like the space of electric charge that we were talking about earlier on. People have found that the old dimensions of twenty-six for the bosonic string and ten for the theory which has fermions are unnecessary. You can formulate theories in dimensions which are less than twenty-six or ten if you trade in some of these spacetime dimensions in a mathematically correct way, which is rather difficult to describe.

That seems to me to be going backwards. One of the attractive features about these recent attempts at unifying the forces of nature is that one is replacing what were previously abstract internal symmetries and properties by concrete geometrical structures, in the form of extra dimensions. Isn't it, in fact a retrogressive step to do away with them?

Maybe it's a little bit emotional to use the word retrogressive. I think we just have to follow wherever the mathematics and the physics goes, and I think more or less by definition it's progressive rather than retrogressive.

But it could very well be that all these string theories in lower dimensions are, in fact, different ways of looking at the original ten- or twenty-six-dimensional theories. It may well be that we are talking about the same thing as before, but just using somewhat different language.

Before we leave the subject of higher dimensions, and their compactification, is it the case that these little loops that we've been talking about can actually loop around these tubes, loop around the rolled up newspaper?

You can imagine some very complicated configurations. If we come back to our newspaper analogy, and imagine having a loop of string, well you can wrap it around the rolled up newspaper once, or you can wrap it around twice, or you can wrap it around three times and so on.

You could put a twist in it as well.

Yes, well, that would be if you had a somewhat more complicated type of string. But yes, you can imagine all sorts of possibilities like that. If you want to make a twist, for example, think not in terms of a piece of string perhaps but an elastic band which you can imagine putting a twist in. So, yes, those are other topological properties which are offered by the theory, although I don't think that we can yet say that we really understand them.

It certainly seems to be the case that theoretical physicists have to dip into branches of mathematics in which they previously may not have taken an interest in in order to get to grips with this superstring theory.

Indeed. I find myself touring through the bookshops trying to find encyclopaedias of mathematics so that I can mug up on all these mathematical concepts like homology and homotopy and all this sort of stuff which I never bothered to learn before!

Could we turn now to the possible experimental tests of the theory because I think we are all agreed that here we have a very exciting and beautiful idea, but of course, ultimately science has to rest on experimentation. What conceivable experimental tests are there for the superstring theory?

I mentioned a while ago that one of the possibilities suggested by the superstring was that there might be two photon-like particles or maybe even three photon-like particles. Now these extra photons couldn't actually be massless particles like the photons which we are using on this radio programme. They would have to be massive particles rather like the W and the Z which were found at CERN a few years ago. But it could very well be that, for example, there would be a second Z-type particle which would have certain characteristic properties suggested by the superstring. One of the things that people are doing at CERN now is looking for possible signs of this extra Z boson.

And are there any other sorts of particles that would be predicted by the theory?

In addition to the known electron-type particles and the corresponding neutrinos and the quarks which I've also mentioned, there could be additional types of matter particle which act in some ways like quarks but in another sense act partially like electrons, so-called leptoquark particles. This possibility is suggested by the superstring, and although we can't be sure that leptoquarks really exist, they are at least something which it seems reasonable to look for experimentally.

What are the chances that we will actually have an experimental handle on this theory in the foreseeable future?

That's very difficult to tell. I don't think we understand the theory well enough to know whether any of these new types of photon or new types of matter particle are really serious predictions of the superstring. Even if we could be convinced that they were serious predictions, we wouldn't know what mass they would have and what energies we would need in our accelerators in order to be able to produce them. At the moment we are just groping around in the dark and having a go to find whether we can reach out and touch anything. It may well be that there's actually nothing there!

Of course one of the things that physicists have traditionally done when faced with this sort of problem is turn to the subject of cosmology for confirmation. Presumably during the very early stages of the universe, in the so-called big bang, there would have been enormous energies released and one might have expected that superstring activity would have left some sort of imprint on the universe that we could see today. Do you think that's the case?

That's certainly possible. For example, one of the things which we believe exists in the universe today is something called 'dark matter'. This is matter which doesn't shine, it doesn't couple to photons, we can't see it with our telescopes. But we know it must be there because we can measure, in a rough way, the gravitational forces which the different particles in the universe exert upon each other, and it seems that out there there must be some sort of hidden dark matter which is doing a bit of gravitational attracting in addition to the stuff that we can see.

What this dark matter might be we don't know, but certainly one of the possibilities is that some relic particles left over from the early stages of the big bang are floating around out there. In the superstring theory it is at least possible that one of the different types of string oscillations, one of these harmonics if you like, might actually be a stable particle which could be a relic left over from the big bang.

Would you expect, if the superstring theory is correct, that the universe may have developed differently in the early stages from the conventional model, that its dynamics would have been modified by the presence of the superstring?

I think that's certainly the case. Imagine going back earlier and earlier towards the beginning of the universe. Then, for example, all the light elements which we know about in the universe, such as helium and deuterium and tritium and so on, those would have been made when the universe was about a hundred seconds old. At that stage presumably, the known laws of physics are entirely adequate to describe what went on. Now if you go back earlier than that then it could very well be that the superstring would make different predictions from the standard model for the early evolution of the universe. I don't think that we yet know the theory well enough to be able to make a precise statement about the way in which it is modified. But certainly one of the things that we would have to take into account would be the possibility that if you went back sufficiently early in the history of the universe, instead of the universe having three space plus one time dimension as it does now, it would look multi-dimensional. Maybe its dimensionality would increase to ten or twenty-six.

In other words, the curling up that we were talking about didn't happen until a little while after the origin of the universe?

That's right. It could very well be that very close to the origin the universe really did have twenty-six or maybe ten dimensions, and then as the universe evolved some of these dimensions, for some reason which we don't fully understand yet, just spontaneously decided to roll themselves up and then the universe carried along with four dimensions of the type which we can see today.

Turning to a more philosophical point, looking at the superstring programme historically, it seems that physicists have stumbled onto this by accident. At present we have a mathematical procedure, albeit rather abstract, but it does seem possible that it's going to describe all of the particles and forces of nature. One is bound to ask why this is the case. Is there some deep underlying principle upon which it is all based or is it just some sort of accident that we happen to have discovered the formula that unlocks the secrets of nature?

Well, I think that it's quite true that the string was discovered more or less by accident, about fifteen or sixteen years ago now. In fact, when it was discovered people didn't think that this was a Theory of Everything. In those days it was imagined as being a possible alternative to quarks for describing nuclear interactions and nuclear forces. Then we found out that string theory really wasn't such a good description of those nuclear forces. We found instead a description in terms of gauge theories, where the fundamental interactions, like electromagnetism, the strong nuclear force or the weak force, are mediated by spin 1 particles rather like photons. For the last fifteen years, the language used for discussing physics has been the language of gauge theories.

Now to come to string theory, we believe that this is something like a sort of superduper gauge theory with an enormous number of symmetries which we are only just beginning to understand. It would include the sort of gauge theory which we have been playing around with for the last fifteen years, but it would include many other things as well. For example, it would include Einsteinian general relativity, as another one of the enormous numbers of special symmetries which it contains. Nowadays string theory would be regarded as a candidate for unifying gravitation (that is, Einstein's theory of general relativity) with the sorts of gauge theories which we have developed over the last fifteen years for the strong, weak and electromagnetic interactions.

This does raise a point that I was going to ask you anyway. I think a lot of nonspecialists might find it a little bit baffling that a theory which is addressing the fundamental particles of subatomic matter, and the forces that act between them, should involve gravitation in such a basic way. Is there some easy way of seeing why gravitation is important to particle physics? Why does gravity have to be in there at all?

Well, we know that elementary particles do have gravitational forces. In fact, they've even been measured in the laboratory. You can slow down an elementary particle until it's moving very, very slowly and then you find that its path bends under the Earth's gravitational force. So certainly elementary particles do have gravitational forces, and if we want to seriously pretend that we have a Theory of Everything, then we are going to have to include gravitational forces in our description of the fundamental interactions.

There is, however, a deeper issue. Ever since the days of Einstein and the quantum revolution, there's been a big puzzle in fundamental physics which has never been resolved, which is how to reconcile gravitation with quantum theory. This is something which many well-known physicists have struggled with in vain, but they have never been able to get a quantum theory of gravity to work correctly. Now it seems that this superstring Theory of Everything may actually be able to do that. At least some of the string theories seem to be theories for which higher order corrections all turn out to be finite, and that's something which is very unusual for a quantum theory. In fact, when people have tried to make quantum theories of gravity before, they have always found that when they tried to calculate something, it turned out to be infinite in an uncontrollable way – a way that they couldn't make any sense of. Well here, touch wood, we seem to have a theory which behaves itself. That's one of the main reasons why people are so excited about the string theory. It may finally be able to reconcile two of the greatest physics revolutions of the twentieth century, namely, quantum mechanics and Einsteinian general relativity.

This is not the first attempt at a completely unified theory of nature – a Theory of Everything. Is it, in fact, going to be the last?

Who knows! You ask me to look into a crystal ball which is distinctly cloudy.

Supposing it doesn't work though. Do you think that this is the last chance to construct a scheme where nature is mirrored in a simple way in bits and pieces of mathematics?

I certainly don't think it's the last chance. With regard to our experiments in particle physics, we can do experiments at energies of the order of 100 GeV, namely about 100 times the mass–energy of a proton. Gravitation is a force which also has an intrinsic energy scale associated with it, the so-called Planck energy. That Planck energy is 10^{19} GeV which is many, many orders of magnitude larger than the energies that we can now achieve in the laboratory.

Presumably somewhere between 100 GeV and 10^{19} GeV we will eventually discover all the elements which go together to make a Theory of Everything. The superstring as it's currently used is a very bold and speculative, some people might even say crazy, proposal which suggests that already with the physics which we can see around the 100 GeV scale, we might be able to jump all the way to 10^{19} GeV. But that's a gamble which may well not pay off. It may well be that we are going to have to work our way laboriously through the physics of 1000 GeV and 10,000 GeV and gradually build up more and more understanding, until we eventually reach the Theory of Everything somewhere off in the distant future.

Even if the string theories which people have constructed up to now do not turn out to be *the* final answer, I think that string theory has arrived as a language to discuss fundamental physics, elementary particle physics and relativity physics. I don't think that there is any danger that in a few year's time we are just going to forget about string theory. I think that even if it doesn't turn out to be the Theory of Everything, or we can't prove it's the Theory of Everything in the near future, it will nevertheless remain part of our fundamental physics vocabulary.

Let us for a moment stand back from the subject and look at it in a sociological context. You have written that the superstring programme has given rise to a 'totalitarian fervour'. It certainly is the case in my experience that the superstring idea has gripped the physics community as no other theory has before. So obviously people's judgements are affected by the excitement, the euphoria. But trying to look at this objectively, what are the major outstanding problems that remain? One of them is the nature of the compactification. Are there any others?

I think there is a question as to whether or not compactification is actually necessary, or whether somehow the theory could be formulated in four dimensions right from the beginning. But if there is compactification then we obviously have to figure out how it takes place, and what determines why one way of rolling up the 'newspaper' is better than some other way of rolling up the 'newspaper' so that we can calculate, for example, the number of electron-type particles and the number of photons. So that's a second very important question.

There are certainly many other types of questions which are very important. For example, we have to understand why all the different elementary particles have the masses that they do. Why some of them are very much lighter than this Planck energy scale of 10^{19} GeV, and where the nonzero masses come from anyway. We believe that the masses come from a mysterious object called the Higgs boson, but we know that in order to make a sensible description of the Higgs boson the theory must be supplemented by something. A lot of us believe that that 'extra something' is supersymmetry, and it's the 'super' prefix in supersymmetry which gives the superstring its name, because the superstring is a version of the string which contains supersymmetry. Supersymmetry seems to be necessary for consistently giving masses to particles.

We have been talking a lot about the four fundamental forces of nature, but in the last year or so there has been some speculation that there might be more than four forces – a so-called fifth force has been discussed. If there is a fifth force, could it be made to fit into this scheme somewhere?

First let me say that I'm not a great believer in the fifth force. I think that the evidence for it is very, very weak, and I, myself, don't take it terribly seriously. Now there are some people who claim that you could find a place for the fifth force within the superstring framework. Again I must say that I'm somewhat sceptical. I think I'd rather sit that one out.

It sounds really rather grand to have a Theory of Everything, and one would certainly be exhilarated if we could write down such a theory and have confidence in it, but would it mean the end of physics? Should physicists then pack up and go and do something else?

I certainly don't think so. In fact, the majority of physicists are not in the business of trying to uncover new laws of nature, they are trying to understand better the way in which nature uses laws which are already known. Most of them are doing that in the context of some sort of laws or, as we would say, a Lagrangian or Hamiltonian, which has been established previously by someone else. It's people in elementary particle physics and in gravitational physics who I think are really involved in finding out new laws of nature. I think that all that would happen if we really did have a Theory of Everything, is that the business of particle physics and relativity physics would become more like that of other branches of physics, for example, solid state or condensed matter physics.

What one might call, I suppose, applied physics.

Well, some of it might be rather applied but some of it is fairly misapplied or unapplied as well, I'm afraid.

Coming back to the question of the experimental verification of the theory, big accelerators are expensive, and we can't expect to see many more, or at least we can't expect to see accelerators in the future that are very much bigger than the ones we have. So it seems to me that the burden for testing the superstring theory (or whatever other Theories of Everything that come along) is going to rest upon foreseeable accelerator designs. You work at the CERN laboratory near Geneva, where one of the prestige projects being set up at the moment is the so-called LEP accelerator. Is there any chance that LEP is going to be able to test some of the ideas we have been talking about? Would it have enough energy to actually probe into the region necessary to verify the superstring theory?

Well, I think there is an outside possibIlity that LEP might be able to produce a second Z particle, although that's probably rather unlikely. Probably what we would have to content ourselves with is indirect evidence for this superstring particle, if we find any at all. One of the things that is planned for the LEP experiments is to look very, very carefully, in great detail, at the properties of the first Z particle. Various proposed detailed tests of the properties of that particle could help tell us whether or not there is another one lurking around somewhere.

Another possibility is that at LEP we can actually count the total number of species of elementary particles in the universe, so, according to the superstring theory, that would enable us to pin down the topology of the compactifying space. It would enable us to tell, in some sense, how the 'newspaper' is rolled up.

Still a third possibility is that certain types of extra particles which exist in some superstring theories, might possibly be produced at LEP or one of the other accelerators now around. For example, there are the leptoquark particles which behave like a combination of a conventional quark and a conventional electron. It's conceivable that some of those could be produced at LEP.

When can we expect to get some results?

Well, LEP is supposed to start doing its experiments in 1989.

In recent months, there has been some debate as to whether or not Britain can afford to remain in the CERN laboratory, as a contributor to this European enterprise. If Britain should pull out of CERN at the time when these exciting ideas are being tested, is that going to be a really major blow for British science?

I think that it would effectively mean that Britain would be withdrawing from this particular type of fundamental science. Remember, science is something where you make theories, but you have to test them with experiment. Withdrawal from CERN would mean that Britain would be essentially cutting itself off from the possibility of doing experiments. Without experiments, I don't think that what you are doing is science.

7

Abdus Salam

Abdus Salam is Director of the International Centre for Theoretical Physics, Trieste, and Professor in the Department of Physics at Imperial College, London. He has contributed to many important advances in particle physics and quantum gravity, and was awarded a Nobel prize for his work on the unification of the weak and electromagnetic forces. In recent years he has turned his attention to superstrings.

A hundred years ago it was widely believed that physics was approaching an end: that Newtonian mechanics, Maxwell's electromagnetism, and other parts of physics, really did describe all of nature and it was just a matter of filling in the last few details. Then it all fell to bits at the turn of the century with what we might call the 'new physics'. But now there seems to be once again this feeling that, although we may perceive it only dimly, there is appearing the outline of a complete Theory of Everything, a theory in which all of nature is encompassed within a single unified description. Is this just a mirage or are we really approaching the culmination of theoretical physics this time? How warmly do you feel towards these new ideas?

If you are asking about the superstring theory and its significance, I feel very warm. However, that we can ever get to a final Theory of Everything, is something that I personally do not believe. After all, no theory should be believed in beyond what one can test. The claim of the present Theory of Everything is that it can tell us about all phenomena up to Planck energies (about 10^{19} GeV). Now, to test a theory *directly* at Planck energy, we would need accelerators delivering such energies. On any forseeable future design such accelerators would have to be at least 10 light years in length! So we can never have a direct test of any Theory of Everything, valid for energies higher than, say, 10^7 GeV. Only indirect tests may be possible, but these can never be all embracing. Superstring theory is exciting because of its own intrinsic merits. At last we have found a real substitute for field theory of point particles. It was this point-particles concept which had been responsible for intractable difficulties in quantum gravity theory in the past.

Why should we want to replace the concept of point particles?

Because this promises – for the first time – a quantum theory of gravity. This is a triumph, irrespective of whether we are getting a final Theory of Everything. That such a theory of gravity also comes unified with a theory of quarks, of gauge particles – the photons, the Ws and the Zs – is an added bounty. But even if this unification had not happened, I would consider strings an important development.

What is the key feature of a string as opposed to a particle that enables you to achieve such promising progress?

We have replaced a point particle with a finite-sized object – a string with a finite size of the order of 10^{-33} centimetres.

The theory of strings provides something which Bohr would have loved – a finite fundamental length of 10^{-33} centimetres. But in spite of this finite length the theory is still local. That is the incredible part of it.

In what sense is it local?

Local in the sense that causality is still preserved. The events which are space-like situated do not disturb each other.

The beauty of the string theory is that, even though we are dealing with an extended object, the string interactions do take place at one point – they do not take place all over the string. The strings split and recombine at just one point in their extension, and when the strings touch each other they touch at a point. This is the secret of their locality.

So we should think of the strings not only as models for particles of matter but also as models for the way in which these particles interact with each other?

Yes. From this point of view, whether strings will explain the whole of physics is a secondary consideration. The strings have been around for more than ten years. But even their most ardent proponents did not emphasize enough this particular merit of theirs – that they could provide a local and causal theory of quantum gravity.

What was the reason that they suddenly became so popular?

The technical advantage of anomaly cancellation could be built into the theory if one unified gravity with a special Yang–Mills theory. This basic discovery made by Green and Schwarz led, on the one hand, to a *unique* unified theory of gravity and of a special set of Yang–Mills gauge particles, while on the other hand, anomaly cancellations made it plausible that this unique theory may also be finite.

We still need to prove rigorously that these theories do give finite results. However, the prospects seem good.

Could you explain what finiteness means?

Most theories of quantum gravity which had been developed in the past gave results which were infinite. Suppose one wanted to compute the scattering of a graviton from another graviton. The answer in pre-string theories of quantum gravity was always: infinity. Such a theory, which gives infinite answers, is worthless and intrinsically inconsistent. The superstring theory is the first theory of quantum gravity which promises, for example, to give a finite result for the scattering of gravitons from a graviton.

Now the interesting thing about string-based quantum gravity theory, is that it is not *gravity* which is emphasized in the string picture.

Here I would like to repeat a remark made by Chris Isham at Imperial College. He said that when he was a student, he started to work in the area of quantum gravity with the hope that by imposing appropriate consistency conditions on the theory of gravity, he might solve the mystery of why gravity *must be quantized*. In other words, he would derive Planck's quantization from Einstein's notions of general covariance. As it turns out in the string theory, it is just the other way round. We are beginning to find that Planck's quantization has to be primarily introduced, while Einstein's gravitation is emerging as a secondary concept. This is happening because of a property with which we endow the strings – the property of scale invariance, something which Einstein's colleague, Herman Weyl, introduced and which Einstein did not like.

I'm sure Einstein would have been very upset.

He would be upset. He gently chided Weyl with having misled everyone with his ideas. Admittedly Weyl's ideas were introduced in the context of four-dimensional space and time and not within a string concept, but Einstein wrote to Weyl saying, 'I'll have to protest to *Der Alte* against your ideas'!

If these superstring theories turn out to be finite, then that would be very compelling. However, it would be nice to have definite new predictions that could be tested, rather than merely reproduce the physics we know. Is there any chance that these new ideas will lead to definite predictions?

Oh yes. There are some predictions. For example, almost all string theories seem to predict a new Z^0 particle. A Z^0 particle, which might behave like a heavy photon, is a great thing once we know its mass. We do not, at present, know what the mass should be, but suppose the mass can be computed and predicted within a string theory – as was the case for the old Z^0 particle of the pre-string days – which you may recall was the signal of unification of electromagnetism with weak nuclear forces. The new Z^0 will now provide a crucial test of string unification of *all* fundamental forces – electroweak, plus strong-nuclear, plus the gravitational.

Is the mass likely to be in an accessible energy range?

That we do not know. A decent theory of masses with predictive power just does not exist.

But in the string context, there is the prediction of hidden matter – a new type of matter. That may yet acquire a life of its own.

These are at present straws in the wind. I would not like to say that these are firm predictions of the type which we achieved with the theory of W and the old Z^0 particles and which were confirmed experimentally by Rubbia. We hope such firm predictions can be made and verified.

You have touched upon a very fascinating aspect of the superstring theory, which is the idea that there is a duplicate universe made up of a double copy of matter. Could you describe that briefly?

This is the idea that there is a duplicate universe which communicates with us only through gravitational forces and no other force. Amazingly this invisible universe should determine the manner in which supersymmetry should be broken in *our* universe. Such a theory would then shed light on the problem of what determines some of the mass differences in the *visible* universe.

In other words, the existence of this other universe shows itself through the masses of elementary particles? And our universe would do the same for the other one I suppose? It is presumably a symmetric arrangement.

I suppose so. I do not think that anyone has speculated in detail on what the other universe is going to look like.

As far as this other universe is concerned, you say that we would interact only through gravity. So we would notice a black hole, for example, from the other universe, but we wouldn't notice if there were atoms of that other universe in this room – they would just pass straight through us?

Large black holes made of the duplicate matter will not be apprehended except through their gravitational effects – like the invisible djinns of the Arabian Nights. The invisible universe presumably has its own 'electrically' charged quarks, its own Ws, its own photons. But such photons do not make this duplicate matter shine, for they interact with nothing of which the visible universe is made. Let me stress however, that superstrings are not unique in postulating this type of gravitating, but otherwise invisible, universe.

Is there a connection between the superstrings and cosmic strings?

There could be. The connection may come from arguing that our universe started with a small size, with small strings, and then as it inflated, the strings were stretched.

Ed Witten has considered these matters more than anyone else has. I would like to hear him on this.

Coming back to the problems facing superstring theory, you have mentioned that the finiteness problem is perhaps the most pressing because if the theory turns out to be finite, then that is a great triumph. Are there any other problems or obstacles to the further development of the theory?

Even if the theory is finite, to compute with it, is such a pain!

That's what we have graduate students for!

No! Graduate students will not attempt that. You see, the superstring theory can be couched in its most transparent version in ten dimensions: nine of space and one of time. The difficulty comes when we wish to couch the theory in the four spacetime dimensions we live in, compactifying the other six dimensions. Maybe we will train computers to do such calculations, but not graduate students!

Isn't there something terribly wrong about a theory which is supposed to encapsulate all of nature being so incredibly complicated that we can't make progress with it? Shouldn't nature be simple?

Nature is simple if we look at it in the right way. For example, I believe that God created just two dimensions – one of space and one of time. What could be simpler? And then at a later epoch, there was a phase transition to four dimensions, plus six internal ones. Two is the heart of the matter in this theory and this number two will not be changed to three.

Can you formulate superstring theory in two dimensions?

Yes. This is where it becomes the simplest, as Polyakov showed. This is how, I believe, the theory starts life – in two dimensions with ten basic fields. We need this number ten to cancel conformal anomalies and thus get rid of at least one type of potential infinity. Parts of these ten fields may manifest themselves as four dimensions of spacetime, with the remaining six dimensions compactifying themselves as internal dimensions – representing electric or nuclear charges. In this picture four-dimensional spacetime will begin at the moment of this phase transition.

How can space jump between these different numbers of dimensions?

These are normal phase transitions – if we can motivate them within the two-dimensional theory. I said 'if' we can motivate them. No-one has done this yet. But that is my dream.

It is said to be a big problem, going from ten to four isn't it?

That is true. That is where the complexity of the universe as given by the string theory resides.

Do you think the problems with changing numbers of dimensions are as great as the problems of proving finiteness?

No. I think the problems of proving finiteness are difficult because of the unfamiliar mathematics of Riemann surfaces. Apparently Teichmuller is a great name in that subject – a mathematician who died in the Second World War.

But it's keeping an army of theoretical physicists busy at the moment?

Not an army. Most people are busy transforming ten to four space-time plus six internal dimensions. That is the simpler task. Many avenues have been found for achieving this – none compellingly elegant. The harder task – the problem of finiteness – involves higher orders of loops in strings. That is not keeping so many people busy, because it is hard.

Could you describe what these little strings are like? Are they closed loops or are they open strings?

The theory which describes gravity must correspond to a closed string. The oscillations of such a string correspond to the physical particles. These particles must have spins of magnitudes of 1, 2, 3, ... The particles of spins 1 and 2 are massless, corresponding to zero frequencies of the string, while higher spin particles must be massive in multiples of Planck's mass, that is, nearly 10^{19} proton masses. The spin 5/2, 7/2,... particles must be massive in units of Planck's mass.

Are there not formidable mathematical consistency problems about formulating theories of objects with spins greater than 2?

This is the great miracle of this theory. It is finite just because of the higher spins. What is so incredible is that this theory is also local.

Is it accurate to think of what we have always envisaged as a particle as being a closed string loop at low energy without any wriggling motion?

No. We are not speaking of an individual particle. The string describes the entire *set* of all objects of higher spin. They all come together.

Would it be correct to say that these loops of string are not looping around in the three-dimensional space that we see, they are looping around in the higher dimensions?

No, that is not correct. The string loop *is* a loop in the four space-time dimensions, with possible twists in the extra six internal dimensions.

Is it possible that our particle accelerators could actually show us particles with a spin greater than 2?

They all have masses of the order of the Planck mass, so they will not be directly accessible in foreseeable experiments.

But if any sort of new particle is detected, then that will be tremendous.

Oh yes. It will be tremendous, for example, if the theory predicts definitely an extra Z^0 particle with an experimentally accessible mass. It would be tremendous indeed.

And if all this happens, and the superstring theory becomes the generally accepted theory of the fundamental matter and forces, what next? You said that actual computations with the theory are horrendously difficult, so what do we do? Do we just stare at the formulae and marvel over them, put them up on our walls and say 'That's a great accomplishment'?

This is always what happens; take Einstein's gravity. After the three famous tests showed Einstein's theory worked better than its rivals, we came to accept its truth. No further calculations were done for a long time, because they were tremendously complicated.

Should one stop at strings? Why don't we go on to consider more degrees of freedom? Membranes, for example.

At the moment we have a negative theorem. It says that you cannot write a conformally invariant theory of higher-dimensional objects like membranes. Thus no good will come out of membranes. It is a sort of negative theorem I personally dislike because of unwritten assumptions which are usually built into proofs of such theorems. But there it is – a challenge.

Clearly string theory is deeply rooted in geometry. I suppose one could say that science started with geometry (if one goes back to ancient Greece). It would be fascinating if, ultimately, we build all the fundamental things in the world out of geometry.

I was talking recently with Christopher Zeeman – the topologist who founded the Warwick Institute of Mathematics. I asked him how he distinguished between geometry and analysis. He said that the mathematicians have a simple test. If a man is balding, he must be an analyst. If he has a lot of hair, he is a geometer!

It does seem that as particle physics develops, more and more abstract structures have to be invoked and more and more obscure branches of mathematics need to be looked at.

I am glad you mentioned this point because this is something else which fascinates me. Res Jost once said, that all the mathematics which a young man needed to learn after quantum theory was invented, was a rudimentary knowledge of Greek and Latin alphabets in order to populate his equations with indices. No longer so! During the last few years we have seen topology, homotopy, cohomology and then the Calabi–Yau spaces, Riemann surfaces, moduli spaces – real and live mathematics – invade physics. The more real mathematics we know, the deeper our insights are likely to be.

I was talking to my collaborator, John Strathdee, the other day. I marvelled at the mathematics which we have to learn today – real mathematics, not *ersatz* mathematics. He said, don't you think it will damage our brains? – just like Bertrand Russell complained in his autobiography that working on the *Principia Mathematica* permanently injured his brain. I was reminded of the Old Father William poem of Lewis Carroll:

You are old, Father William, the young man said,
And your hair has become very white
And yet you insistently stand on your head
Do you think at your age it is right?

Reply:

In my youth, Father William replied to his son,
I feared it may injure my brain
But now that I am sure I have none,
I do it again and again.

8

Sheldon Glashow

Sheldon Glashow is Higgins Professor at Harvard University, and is also affiliated with Boston University and the University of Houston. He has made important contributions to many aspects of the theory of particle physics, and was awarded a Nobel prize for his foundational work on the unification of the weak and electromagnetic forces. He is also actively concerned with science education. Glashow is an implacable opponent of superstrings, on both scientific and philosophical grounds, and says that he is 'waiting for the superstring to break'.

Can I begin by saying that a hundred years ago it was widely believed that physics was coming to an end, that it was just a matter of dotting a few 'i's and crossing a few 't's. Recently this view seems to have surfaced again. A number of people are talking about the culmination of theoretical physics, about a complete Theory of Everything in nature. Do you think that this is another false alarm?

It's certainly not true that theoretical physics is coming to an end. For example, new discoveries are happening very rapidly and very excitingly in what is called condensed matter physics. I think you are talking about elementary particle physics more than physics as a whole. Particle physicists are, at the moment, at a terribly exciting stage of their subject because they are joining hands with their friends the cosmologists. For once we have a theory that deals both with the microscopic world – the world of high energies and small distances – and also the birth of the universe and the origin of the world as we know it. So this new unity between cosmology and elementary particle physics – signified incidentally at the American laboratory Fermilab, which has a large contingent of astrophysicists in its group – indicates a renaissance, not impending death.

But is the hope that for the first time we might really be able to write down a complete theory of all of nature at a fundamental level – a theory of all the forces and all the particles – just a delusion?

So far we claim only to be able to write down a complete theory of all of the elementary particle forces, that is to say, nuclear forces and electromagnetism but not gravity. The theory we obtain is contrived and *ad hoc* and it has a lot of built-in mysteries. For example, why are the ratios of the particle masses exactly what they are?

We do not yet have a theory which incorporates gravity. We may have the beginnings of such a theory, but just the very beginnings. My friends the string theorists, who work with today's vision of a truly unified theory, including gravity, say that it will take them twenty years to begin to make contact between the world of gravity and the world of elementary particle physics.

They do seem very confident that they really have grasped the essence of a truly unified theory.

They have the feeling that they require, as Ed Witten says, the construction of five new fields of mathematics before they have any reason to become confident that they have a theory. In fact they do not have a theory. They have a complex of ideas which do not evidently form any kind of theory and they cannot even say whether their structure describes the successful accomplishments that have been obtained in the laboratory, and in theoretical physics.

What's the reason for their optimism do you think?

They feel that for the first time they have a consistent quantum theory of gravity, and perhaps they are convinced that they have the *unique* consistent quantum theory of gravity. It may or may not be true. It has a possibility of being true, and for once they think Einstein's dream has a chance of being realised. I always like to remember that Einstein, in his last three decades of life, followed this dream and seemed to be completely unaware of the exciting developments that were happening in nuclear physics during this time.

You once said in a lecture that physicists seem to be dividing into two camps: the alchemists and the medieval theologists. What do you mean by this?

I'm particularly annoyed with my friends, the string theorists, because they cannot say anything about the physical world. Some of them are convinced in the uniqueness and beauty, and therefore truth, of their theory, and since it is unique and true it obviously includes a description of the entire physical world. It does not seem to them to be necessary to do any experiments to prove such a self-evident truth, so they begin to attack the value of experiments from this end – a highly theoretical, abstract, mathematical end – whereas some of our friends in Britain are attacking physics from the other end, from the purely financial end.

So you see this move towards theories that attempt to unify all of nature in this very abstract way as actually threatening the future of physics because they undermine the motivation for experiment?

Yes, in the same way that I think medieval theology destroyed science in Europe in the middle ages. It was, after all, only in Europe that people did not see the great supernova of 1054, for they were too busy arguing how many angels could dance on the head of a pin!

But it is true though, isn't it, that whatever one thinks about the details of these theories, a lot of very interesting physics lies at energies that are way beyond those we could hope to explore directly?

We don't know. It's not clear. Some visions have it that there are no interesting particles remaining to be discovered at these energies, that there is a desert, free of any particles. Other theories claim that the desert is populated throughout with new discoveries yet to be made. I don't know what these string people believe, I don't think they know what to believe since they cannot make contact with low energies, and do not know whether the desert blooms or does not bloom. In any case it hardly matters to them since the theory, if properly developed, would presumably explain it, whatever it may be.

Even if there is something to be found in the desert, to understand all of physics, to really grasp this idea of a complete unification, we would have to go right up to the Planck energy, right up to these very extreme energies.

So the party line goes, but I don't know if that party line is true. How relevant the Planck energy is to elementary particle physics has not really been established, after all. It's merely a number with the dimensions of a mass that comes out of Newton's theory of gravity. Call it the Planck mass if you wish. It may or may not play a fundamental role.

Of course, one could take entirely the opposite point of view from the one you have been advocating and say that this drive towards unity is very compelling, very beautiful, very inspiring, and that it might well stimulate further work in experimental particle physics rather than hinder it. Don't you think there is something about building a complete theory of the world that is going to persuade the people who are going to have to pay for these things, that they should come up with the money to help test these ideas?

Yes, if our superstring ideologues would say that it is necessary to go to higher energies in a realistic fashion, energies that we can afford and can reach. If they would argue that we needed experiments I would strongly agree with them. This is not the sort of argument that many of them seem to make. Many of them seem convinced in some abstract way that it's good to build larger accelerators, in the same fashion that it's good to work on a cure for cancer, but it has nothing to do with the work that they do. They are not a force for driving towards higher energy. It's our experimental friends who want to see more of the universe that are the real driving force of physics, and always have been.

Who's going to win?

I should hope that it will be the experimenters. I think that the old tradition of learning about the world by looking at the world will survive, and we will not succeed in solving the problems of elementary particle physics by the power of pure thought itself.

Now one of the problems about doing experimental physics, is that there have been a remarkable number of false alarms, 'discoveries' of apparently spectacular things which went away again. Do you think experimenters are getting a bit incautious in the way they announce their results?

Experimenters have always been incautious. I am sure there were as many false discoveries in the past as there are false discoveries now. It's just that today they stand out a bit more because there aren't any *real* discoveries to point to, at least in the last five to ten years. It's not a question of carelessness. It's a question of perhaps having the financial and moral support of the nations that affect today's experimenters.

You have spoken about the increasing cost of new accelerators. It's true that each new accelerator costs more money than an accelerator of the past. However, we have many fewer accelerators today. Whereas a few decades ago there were thirty large accelerators in the United States, today there are only three, and we are concentrating our funds on a smaller and smaller number of facilities. Perhaps some day there will only be one giant world facility. But the amount of money being spent is actually decreasing in my country, and I think in Europe as well. As time goes on, in constant units of currency the amount of money being spent on particle physics is decreasing. Consequently there is beginning to be a loss of morale. I hope this will change with the deployment of LEP at CERN, and the surprising discoveries it will almost certainly generate.

You have also referred to the very high rate of genuine discoveries being made in areas like condensed matter physics, which takes only a very tiny fraction of the budget. Now there is a big debate in Britain at the moment about how the budget should be shared among the big spenders – like particle physicists and astronomers – and those working in these other areas which may be potentially more useful to society. How do you feel about the lion's share going to particle physics? Although it's decreasing, it's still enormous isn't it?

It's not at all clear that the lion's share of anything goes to particle physics. It's very complicated how one does the arithmetic. For example, the amount of money spent in my country in the biological sciences in the form of research is much higher than the amount of money being spent on physics as a whole, by something like an order of magnitude. I am sure the same is true in Britain. Now, it is probably true that solid state physics is not adequately supported in Britain, but the reasons might be other than those that appear at first sight. I do not think that the expenditure at CERN, and the roughly equal expenditure on elementary particle research within the UK, represent an undue burden on Britain. Britain is saying it cannot afford to do it. Italy on the contrary has just doubled its budget for elementary particle physics. Is it that England is so much a poorer and meaner nation than Italy?

Coming back to the actual physics, what do you think are the outstanding problems at the moment facing experimental particle physicists?

There are lots of them. One is money, as we have mentioned. Another is the timeliness of the appearance of new facilities. In Europe we are waiting for LEP to be developed, and it's proceeding as fast as can reasonably be expected, but it still takes something like ten years between the conception of a new machine and its deployment. It's still a few years away. It's hard to motivate our young experimenters. It's hard to be sure that by the time the machine works, there'll be a crop of nonsenile experimenters ready to do experiments on LEP.

Another problem is the size of the groups that work at new facilities. Again at LEP, one of the experimental groups contains over 400 PhDs! Now can that work? Can a group of over 400 scientists function as Michael Faraday once functioned years ago? I don't know. Certainly it's a different mode. Can the smart ones be identified? Can the really major contributors become trained and emerge from such groups as scientists in their own right? So far, our experience at CERN is positive. So far the answer, surprisingly, is yes, and I hope that continues.

Looking back over the past ten years or so in particle physics, it seems that there has been a rather remarkable lack of surprises. Doesn't this perhaps indicate that particle physics is beginning to reach the end of the road? Should we really be spending more money on looking for new things in the next ten years, given this record?

I certainly hope that we shall continue exploring. If there really is a desert out there, the only way to be really sure that it exists is to walk through the first few kilometres of sand. It's true that many of the predictions or expectations of the more dramatic theories have not panned out. The grand unified theories predicted proton decay for example, and proton decay was not seen. The theory predicted the possible existence of magnetic monopoles and magnetic monopoles have not been seen. And yet this is the very theory that predicted that these would be the only manifestations available to us at high energy. The fact that these predictions didn't work means, perhaps, that these original naive theories (I can call them naive since I was partially involved with them) are wrong, and there isn't a desert, and there are lots of interesting things remaining to be discovered.

For example, one of the fascinating survivals among the anomalies of today is the problem with neutrinos. We all know that there just aren't enough neutrinos coming from the Sun to agree with our theory of the Sun. We would like to confirm that by studying a wider spectrum of neutrinos. We know that this can be done relatively effectively if one had thirty tonnes of gallium, for example. This experiment is now being done in Italy and in Russia as well. Solar neutrinos are also being studied by another method in Japan. These experiments will tell us whether there is something terribly wrong with our theory of the Sun, or on the other hand whether neutrinos have mass and are subject to oscillations. Either way, this would represent a spectacular new development in particle physics.

The experimenters in quite another field, who explore the structure of the Sun, are learning more and more. They are studying the seismic vibrations of the Sun, and they are discovering that the Sun should be producing even more neutrinos than the most naive theories of the Sun expect. So something is very, very rotten in this whole field. That is to say there is a surprise looming and the subject has to be understood.

In a different direction is a fantastic discovery that has recently been made. You said there were no surprises but there actually have been some. There was the surprise that most of the matter in the universe is invisible, which has just emerged from the astronomers in the last five or six years, much to their embarrassment. They are the people who thought that they were studying the matter of the universe – all of it – but they have discovered that they are, in fact, just looking at that contamination of the universe which happens to emit light, for bizarre reasons. Most of the matter is invisible. What is the form of this matter? This is a problem for astronomers, and it is also a problem for particle physicists and for experimenters. Maybe we can detect this crazy stuff in the laboratory, on Earth, and find out what it is.

Looking forward to the next ten or twenty years, assuming the so-called Superconducting Super-Collider (SSC) machine is built in the United States, which will achieve energies that were unthinkable a few years ago, what will be the priority experiments? What will be the things to go for, the things to look for?

The wonderful thing about surprises is that we don't know quite what they will be, so all we can do is try to design the best wide-ranging experiment that can be done. Of course, we in the States will simply be repeating what the Europeans have done twice, but at higher energy. We are colliding protons against antiprotons, or perhaps protons against protons, at tremendous energies, in the same fashion as the CERN collider now does, except at an energy almost a hundred times larger than the energy of the CERN collider. In broad outline, you allow the protons to collide with antiprotons moving in the other direction and surround the collision area with an elaborate detector involving the newest of solid state technologies. Incidentally, here's an area where there is a lot of feedback between condensed matter physics and particle physics. We build the best detector, which costs ten percent as much as the accelerator itself, and sit back and see what surprises us.

But there must be some things which you would expect to see? What are the things that current theory would lead you to expect to see, that can't be seen with present technology?

Standard theory tells us that we shall see standard things. We shall see the sorts of jets and other curious phenomena that we see at lower energies, simply extrapolated to higher energies. We will be able to test the theory of quantum chromodynamics and the electroweak theory better. But our standard theory predicts standard results. What the 'real' theory – the theory that is not today's standard theory – says, we simply don't know. There'll be new forces perhaps, there'll be new particles, there'll be things that have been variously called glints, or other strange names, by some of my theoretical colleagues. We can't predict exactly what they will be. Perhaps they'll be jets that involve anomalous imbalance of momenta, perhaps there will be single leptons coming out of collisions that have no rational explanation in terms of the standard theory or perhaps we will produce new long-lived particles that are not part of our present philosophy. Or perhaps there'll be something else that I can't tell you, because it is, after all, a surprise. That's the name of the game.

What about the so-called Higgs particle? That's a very important particle to find, isn't it?

Now that's a toughie. The Higgs particle might show up at LEP. Remember that's a big machine that hasn't been exploited yet; it hasn't been built yet. It has a good chance of finding the Higgs particle in a certain range of possible masses. The SSC will also have a good shot at it in a different range of masses where it can decay into two Ws or two Zs. Yes, there is a very good chance of finding the Higgs boson, either at LEP if it's relatively light, or at the SSC if it's relatively heavy. Incidentally, that's one of the things that is not predicted by the standard theory – the mass of the Higgs boson.

What about supersymmetry? Do you think there is any chance that these new machines will reveal supersymmetry? After all, this is a marvellous idea which has been around for a long time, but at the moment there is not a shred of evidence that the world is supersymmetric. When are we going to see some supersymmetric partners of known particles?

Supersymmetry is a lot of fun, and you will recall that not so long ago, when there were some severe anomalies in the data coming out of CERN, data that could not be explained by the standard theory, a lot of people jumped at the chance of explaining these anomalies in terms of supersymmetry. Three different varieties of supersymmetric theory were published trying to explain this. Now it turns out that the anomalies at CERN have all been retracted and do not exist at all, but the same sort of thing could happen at higher energy. Perhaps they will happen at CERN itself when we have more data. At any point it could turn out that supersymmetry is, after all, not only pretty, but right. Then there is technicolour, another kookie theory; also the so-called composite theories, which say that quarks are made of other things. Any of these things could turn out to be right after all, and thereby surprising.

There is, in addition to the techniques of smashing particles together at very high energies, the possibility of exploring high energy phenomena in a different way. I'm thinking of phenomena like proton decay, or looking for relics from the early universe. These are low-energy and presumably low-budget experiments which will nevertheless implicitly explore the physics of very high energies. Do you think that the era of that type of experiment is now over with the failure of proton decay?

No. The experiments that have been done on proton decay are, as you say, relatively inexpensive experiments. It's clear that these experiments can be improved by making them into relatively expensive experiments. It is for this reason that the Japanese wish to build a larger proton decay device, one twenty-two times larger than the one they presently have available, and that will be an expensive experiment – comparable to experiments performed at large accelerators. It is also true that the Italians have built an enormous laboratory under the ground, for the purpose of doing exactly the sorts of experiments that you describe. Some of the experiments there will be very expensive, they will cost fifty million pounds, or something like that. That's not a cheap experiment.

Do you think there's room though, for any completely different designs of nonaccelerator experiments that can probe high energies?

I think we need a little bit of everything. By the way, the proton decay experiments turned out to have a very interesting spin-off. The Japanese developed a twenty inch phototube purely for the purpose of doing the proton decay experiments and have discovered to their delight that there is a real commercial market for this device. They make it. EMI in Britain doesn't. On the other hand, you people have not done the proton decay experiment, therefore you don't have the twenty inch tube. I think we need all kinds of research in this direction.

Here's a real example of serendipity. Neither the Japanese nor the Americans succeeded in finding proton decay. But, because these detectors had been deployed, they were able to detect the neutrinos coming from the new supernova, thus confirming the theoretical speculations of astrophysicists and producing new limits on neutrino masses. Surprises can come from any direction!

There's another interesting spin-off which comes to mind. The SSC that is going to be built in my country will involve a very long tunnel. Therefore there is a lot of work going on now in the States on tunnel technology, on how to build big tunnels cheaply. With that kind of technology, perhaps we could build at half the price a tunnel between France and Britain.

An interesting thought! Returning to the topic of superstrings, how do you see the subject developing?

I'm very happy that so many of my colleagues are working on string theories, because it really keeps them effectively out of my hair. I know that they are not going to say anything about the physical world that I know and I love. Mainly that's the reason that I don't like these theories. I have the greatest respect for the people in Britain and the States who do work on them. At the same time I do everything I can in my power to keep this contagious disease – I should say far more contagious than AIDS – out of Harvard, but so far I've not been very successful. Nonetheless, some of us at Harvard are still trying to follow the upward path, to go from experiment to theory, rather than pursuing the superstring vision, which requires the highest inaccessible dream-like energies to build a theory that deals with the down-to-earth world under our feet.

Given the current favour surrounding string theories, do you think that there is a shift of style in the way in which physics is being conducted these days, compared with fifty years ago?

Absolutely not. There have always been kookie fanatics following strange visions. One of the most kookie, and of course most brilliant, was Einstein himself. It has often been said by my string theory friends that superstrings are going to dominate physics for the next half of a century. Ed Witten has said that. I would like to modify that remark. I would say that string theory will dominate the next fifty years of physics in the same way that Kaluza–Klein theory, another kookie theory upon which string theory is based, has dominated particle physics in the past fifty years. Which is to say, not at all.

9

Richard Feynman

Richard Feynman was Professor in the Department of Physics at the California Institute of Technology. He is credited with laying the theoretical foundations for much of modern particle physics and quantum field theory, and was awarded a Nobel prize for his work on quantum electrodynamics. As one of the 'elder statesmen' of modern fundamental physics, his scepticism towards superstrings is especially pertinent. He died in early 1988.

A few years ago, Stephen Hawking said that he thought the end of theoretical physics might be in sight. I think he was referring to the recent successes in attempting to unify all of physics into a single descriptive scheme. This seems a very provocative statement. How do you feel about this, having spent a lifetime in attempting to unify certain aspects of physics?

I've had a lifetime of that, and I've had a lifetime of people who believe that the answer is just around the corner. But again and again it's been a failure. Eddington, who thought that with the theory of electrons and quantum mechanics everything was going to be simple and then guessed everything, because it was going to be simple, but guessed wrong. Einstein, who thought that he had a unified theory just around the corner, but didn't know anything about nuclei and was unable of course to guess it. And today, there are a large number of things that are not understood. That isn't fully appreciated, and people think they're very close to the answer, but I don't think so.

Do you think we've got any right to suppose that nature at its deepest level is unified – that there are simple mathematical statements that can encapsulate all of reality?

In our field we have the right to do anything we want. It's just a guess. If you guess that everything can be encapsulated in a very small number of laws, you have the right to try. We don't have anything to fear, because if something's wrong we check it against experiment, and experiment may tell us that it's not true. So we can try anything we want. There's no danger in making such a guess. There may be a psychological danger if you bend too much work in the wrong direction, but usually it's not a matter of right and wrong. Whether or not nature has an ultimate, simple, unified, beautiful form is an open question, and I don't want to say either way. I'll find out – although maybe I won't live long enough. I want to find out as much as I can about nature, and not to presume ahead of time. Whether it's a simple formula, or it's complicated, doesn't make any difference. Everyone's guesses can turn out the way they want.

One of the problems about testing these recent ideas experimentally is that the features which suggest there might be unification occur only at exceedingly high energies. I think we're beginning to reach the end of the road for high energy particle physics as far as accelerators are concerned. It's hard to see beyond the next generation of experiments, simply because of financial constraints. Do you think for these reasons that maybe theoretical physics is degenerating into philosophy?

Maybe theoretical physics is degenerating but I don't know into what. Let me just say something first. I have noticed when I was younger, that lots of old men in the field couldn't understand new ideas very well, and resisted them with one method or another, and that they were very foolish in saying these ideas were wrong – such as Einstein not being able to take quantum mechanics. I'm an old man now, and these are new ideas, and they look crazy to me, and they look like they're on the wrong track. Now I know that other old men have been very foolish in saying things like this, and, therefore, I would be very foolish to say this is nonsense. I am going to be very foolish, because I do feel strongly that this is nonsense! I can't help it, even though I know the danger in such a point of view. So perhaps I could entertain future historians by saying I think all this superstring stuff is crazy and is in the wrong direction.

What is it you don't like about it?

I don't like that they're not calculating anything. I don't like that they don't check their ideas. I don't like that for anything that disagrees with an experiment, they cook up an explanation – a fix-up to say 'Well, it still might be true'. For example, the theory requires ten dimensions. Well, maybe there's a way of wrapping up six of the dimensions. Yes, that's possible mathematically, but why not seven? When they write their equation, the *equation* should decide how many of these things get wrapped up, *not* the desire to agree with experiment. In other words, there's no reason whatsoever in superstring theory that it isn't eight of the ten dimensions that get wrapped up and that the result is only two dimensions, which would be completely in disagreement with experience. So the fact that it might disagree with experience is very tenuous, it doesn't produce anything; it has to be excused most of the time. It doesn't look right.

Is this a problem of style of research, or is it a problem of the sort of things that these people are trying to do?

No, I don't know whether you'd call it style of research, it's a question of verifying your ideas against experiment and whether the theory is precise enough. It is precise mathematically, but the mathematics is far too difficult for the individuals who are doing it, and they don't draw their conclusions with any rigour. So they just guess.

You seem to be saying that they're careless in the way they're going about it.

No, there're not careless, but it's very difficult. So they're unable to make a precise prediction – not through carelessness, but through inability. But they then continue to say it looks like a promising theory, in spite of the fact that they have to add all these guesses. Now it could be that it wraps up six of the ten dimensions, and it could be that it does this and it does that. For instance, there's a large number of particles in this theory, many more than we see. So we say right, the ones that we don't see maybe acquire an enormous mass – the so-called Planck mass – which has prevented us from seeing them. And the ones that we do see don't acquire such an enormous mass. But why is it those and not the others? The answer should be a consequence of the very theory they're writing down. But they are unable to demonstrate this. In other words, there's no real comparison to experiment. On top of that, the particles that we see do acquire a mass, but it's much smaller than the Planck mass – it's at the range of present experiment. And how that comes about, this other scale of mass, is not known.

Finally, although people say that there are no experiments to lead us, it's not true. We have some twenty-four or more – I don't know the exact number – mysterious numbers associated with masses. Why is it that the mass of the muon compared with the electron is exactly 206 or whatever it is, why are the masses of the various particles such as quarks what they are? All these numbers, and others analogous to that – which amount to some two dozen – have *no* explanations in these string theories – absolutely none! There's not an idea at the present time, in any of the theoretical structures that I have heard of, which will give a clue as to why those masses are what they are.

So we have a large number of experimental facts already collected, for which there hasn't been the imagination to produce a reasonable theory of any of them. That's where we should start working. That's where we have our real problem, because we have experimental numbers to check against; we could get rid of any theory that you could make up easily by that data. So far, there haven't been any good theories. When you look at these numbers, they look absolutely at random and hectic; there doesn't seem to be much pattern in them. That's a problem for theoretical physics, and these superstring theories don't address it at all.

My impression about these sorts of projects is that they're founded upon broad concepts. In this case the concept that there is some simple elegant piece of mathematics that would draw everything together, but which is only manifest in a regime that may be forever unobservable. Only afterwards does one worry about the low energy limit of the theory, and try to fit these numbers, and that's very messy and technical. Do you believe that this sort of philosophical approach – the idea of some grand fundamental principle underlying everything – is a good one to inspire physicists? It clearly does inspire a number of physicists, but is it maybe misleading that we should approach physics in this way?

I have already answered that question – you can do whatever you want. The only thing that's dangerous is that everybody does the same thing! It may be that there is some wonderful unifying principle, that, in fact, the thing they're guessing is *right*. That's fine if we can demonstrate it. But there are maybe other possibilities. The mere statement that there should be some kind of unification, does not indicate *what* kind of unification. There are enormous numbers of possibilities. Any one of these might be right, or none! And we have to explore. So we ought to be running around in as many directions as possible.

What about the idea of using strings, as opposed to particles as the fundamental objects. Don't you think that idea has any attractive features?

Not particularly. No, the question isn't one idea, or another, or who is attracted to the idea – the question is to get a lot of variety of ideas and to bring them to a point where we can get rid of them by observations. A friend of mine once said to me – we were undergraduate students at MIT – he said 'I think I understand that the problem in theoretical physics is to prove yourself wrong as quickly as possible!' Now what the string theorists are doing is they aren't proving themselves wrong, because they're allowing themselves liberties with their equations and saying 'Okay, maybe it wraps up six of the ten dimensions and leaves us four', without proving that it wraps up six or without considering why it doesn't wrap up seven. They're not checking the ideas hard enough against experiment because of the difficulty in calculating anything. That means they're up in the air and I don't have to pay much attention!

A lot of people who work on superstrings believe that one of the main reasons for studying them is that they promise to dispose of this problem of the infinities, the divergencies, that have plagued fundamental physics for decades now. I would have thought that you should perhaps welcome these theories as solving once and for all this problem of the infinities.

Well, we welcome something or not welcome something depending on whether it's going to agree with the phenomena of nature. Of course it's delightful if superstring theory does indeed remove the infinities. However, my feeling has been – and I could be quite wrong – that there must be more than one way to skin a cat. I don't think there's only one way to get rid of the infinities. I don't think the statement that the theories can't have any infinities leads us uniquely to this string theory. It should lead us in all directions, and because the imagination of man is very high, he will find a whole lot of other ways of getting rid of the infinities, any one of which may be the right theory. The fact that a theory gets rid of the infinities is to *me* not a sufficient reason for believing in its uniqueness. That's my opinion; it's perhaps an incorrect opinion, as I tried to explain – I'm an old man. Maybe these guys understand better than I do that there isn't any other way to go. If I studied it better perhaps I'd understand also why it has to go in this direction. I don't see that though.

But these infinities have been so hard to get rid of, that if the superstring theory really did turn out to be finite, this would be a very compelling reason for believing the theory.

If it also agreed with experiment, yes. But what they say is 'Suppose we take the view that there's no way to get rid of the infinities and then we suddenly discover there is one way to get rid of the infinities, but you can't work out its consequences. Since it's so compelling, it must be the right theory'. And so you sit around saying 'Well, you see, you can't disprove it'. That I can understand. You've explained to me what all these guys are saying and how they could be saying all these things when I didn't understand them. They're not deducing anything, they're just showing that since this is the only model they can make and since this cannot be disproven it must be true. Okay. It's possibly what drives them on. They may be right. I don't think so though!

If we look back to the time when you developed QED, quantum electrodynamics, infinities were, of course, a problem then. In a sense the problem went away because you managed to fix up the difficulties with the infinities by packaging them up and sweeping them on one side, if I can put it that way.

Exactly, yes. That's quite accurate.

So these infinities have plagued quantum field theory for over a generation. Do you think that a fundamental theory of different particle interactions can still contain these infinities? Or do you think that Dirac was right to say he couldn't believe any theory that contained these infinities?

Well, obviously there are no infinities in observation – the mass of the electron is not infinite. When we take electrodynamics in a conventional sense, without adding all the new modifications, we write the equation down and then calculate the mass in the electron and find it to be infinite. Then we have to play a kind of shell game and say that this isn't the way we're supposed to calculate the mass. We're supposed to substract something from something and do this and that and these rules which are called 'the renormalization' or 're-organizational' rules will produce the theory in which all the answers are finite and agree with experiment. That seems to be the case, but we do not know whether this re-organized form is a mathematically consistent form. What's very interesting is that, in all these years, we have never proven one way or the other whether it is consistent. But, for the moment, suppose that it turns out that it *is* consistent. Then we have a mathematical structure which is 'write these wrong equations', that is, when you get infinities play this subtraction game invented by these three guys back in 1947, and take the limit and straighten it out and this will be a finite theory and those are the answers. This is a mathematical structure, even though it sounds messy. Now it should be possible one day for someone to work out more carefully in a different way a set of equations in which there aren't any infinities and which have the same consequences. I don't mean by inventing a new physics, but rather by reorganising the statement of what it is you do to make the calculations less awkwardly written.

So it's just a matter of mathematical technology in that case. But it's also possible that electrodynamics is not a consistent theory, in which case the problem would be much more serious from a physical standpoint. If we didn't have a mathematically consistent theory, we would have to learn more about nature and find out what modifications to electrodynamics would be necessary. We do have some clues to this dichotomy. In an analogous theory called 'quantum chromodynamics' which involves quarks and gluons, and which is supposed to explain the properties of protons and so on, we have a theory which we can prove to be mathematically consistent. It has infinities and they can be swept under the rug in the usual way. The final result, however, is known to be mathematically consistent. It therefore must be possible to say what the result is without going through the infinities. So I think that those infinities are somehow technical. We're formulating the theories incorrectly when we first write them down.

Of course, it's fashionable to suppose that the problems of the infinities will only be solved within the context of a unification of the different forces.

Yes, because of the apparent solution, in the case of quantum chromodynamics, and assuming that the electrodynamics can be proved to be unsatisfactory then in order to get it satisfactory, it's got to be part of an analogous theory. This would mean adopting a certain kind of expansion and increased symmetry with all the different kinds of forces involved in the same packet. That's one of the ideas that has suggested these unified theories. It's a powerful suggestion. I must say right away that I never thought trying to get rid of the infinities would be a good way to discover correct physical laws, and I was wrong. I've often been wrong in guessing the best way to proceed.

When you asked me in the beginning about these superstrings, my modesty is a result of experience. I am unable to say for sure – I just don't believe it. I have believed before that some theories were not going to be any good and they turned out to be good. So, I've been wrong before!

Of course, the really tough problem as far as the infinities are concerned is gravity. It seems almost inevitable that in any unified description of the fundamental forces, gravity is going to play a central role. It may seem odd to some people that gravity should in any way be involved in particle physics because it's such a weak force on the atomic scale. Is there a simple way do you think of seeing why gravity is relevant to these issues?

Well, I'm surprised to hear that you think that gravity might not be important. It's one of the laws of physics! It's obvious that when large masses of material get together, they attract each other. If we are to get a theory of the physical world, and then not be able to explain why large masses come together, we obviously haven't got the right description of the world! So gravity must come out of whatever laws we propose.

But do you think we need gravity to fix up particle physics?

Fix up in what respect?

To solve the problem of the divergences.

Oh, I haven't the slightest idea. It's possible, but the reason we need gravity is because gravity's there. We have to have a theory that explains what we see. That's why we have to have gravity, and never mind whether we need it to straighten out some infinities. The next question is whether gravity has to be a quantum mechanical theory, like the quantum mechanical phenomena associated with the other particles. It doesn't seem possible to have the world partly classical and partly quantum mechanical. Therefore, for example, the fact that you can't observe a position and a momentum at the same time with arbitrary accuracy – which is what we know from quantum mechanics – should apply to gravity also. We shouldn't be able to use gravitational forces to determine the position and momentum of a particle beyond a certain accuracy, because we'd run into an inconsistency. In trying to modify gravity theory to make it into a quantum theory we discover infinities just like we did in electrodynamics, but which are much more difficult to sweep under the rug. They're much more serious. I don't know how gravity fits into these things, but it has to fit in. It presents a very large number of problems beside the infinities.

In the quantum field theories, there is an energy associated with what we call the vacuum in which everything has settled down to the lowest energy; that energy is not zero – according to the theory. Now gravity is supposed to interact with every form of energy and should interact then with this vacuum energy. And therefore, so to speak, a vacuum would have a weight – an equivalent mass energy – and would produce a gravitational field. Well, it doesn't! The gravitational field produced by the energy in the electromagnetic field in a vacuum – where there's no light, just quiet, nothing – should be enormous, so enormous, it would be obvious. The fact is, it's zero! Or so small that it's completely in disagreement with what we'd expect from the field theory. This problem is sometimes called the cosmological constant problem. It suggests that we're missing something in our formulation of the theory of gravity. It's even possible that the cause of the trouble – the infinities – arises from the gravity interacting with its own energy in a vacuum. And we started off wrong because we already know there's something wrong with the idea that gravity should interact with the energy of a vacuum. So I think the first thing we should understand is how to formulate gravity so that it doesn't interact with the energy in a vacuum. Or maybe we need to formulate the field theories so there isn't any energy in a vacuum in the first place. In other words, there are some mysteries associated with the problem of quantizing gravity which go beyond the infinities. They have to do with the formulation of the theory in the first place.

There are also some conceptual issues. If you're applying quantum mechanics to gravity, then in a sense you're applying quantum mechanics to space and time. Now if we take the whole of spacetime we've got the whole universe. It's fashionable these days to talk about quantum cosmology, in which one attempts to apply the laws of quantum mechanics to some simplified version of the universe as a whole. Do you think that the conceptual issues here are really fundamental, or are they just incidental? In other words, do we really have to understand what we mean by, say, the (quantum mechanical) wave function before we can make progress in quantizing gravity?

Only after we make progress will we learn what it is we have to understand and what concepts are unnecessary. It's not easy to tell ahead of time.

Many people who work in this area subscribe to the so-called 'many universes interpretation' of quantum mechanics. What is your feeling about that interpretation?

I don't know. You see we in this field have a tremendous advantage over people in some other fields because we experiment to check our ideas. It therefore doesn't make any difference what you think except psychologically. If you say 'Infinities are impossible, I'll have to make a new theory of this kind' then you could be dead wrong; but you try to make a new theory and it agrees with experiment even though the idea that got you to think of the new thing may not be right. The fact that the new theory agrees with experiment is fine, and you've discovered something. These thoughts ahead of time about what's philosophically consistent and what's philosophically necessary are psychological pushes that say 'I can't believe the theory of today because blah blah, and I've got to go off and try to find something else'. Just like when I was young I said I can't believe in an electron acting on itself; I've gotta find something else. Well I went off; I didn't find the right thing, but I could have. That doesn't mean that it's really true that electrons can't act upon themselves; it's just that the idea was a useful psychological driver to produce a new theory.

So I don't object and I don't want to argue with people who insist that such and such is impossible and this must be so and so. I'm going to work on trying to find a theory which has this new property, because that theory might be right. Okay? I don't want to get into these tangles, because I don't want to discourage any reasonable idea that people have about how things have to be, because it might make them think of something that works. It doesn't have to be true, it just has to make them think of something that works.

So you take a very pragmatic view of these things?

Yes I suppose you'd call it pragmatic, in the sense that all I'm interested in is trying to find a set of rules which would agree with the behaviour of nature, and not try to go very far beyond that. I find most philosophical discussions are psychologically useful but, in the end, when you look back historically at what was being said, and being said with such vigour, it's almost always – to a degree – nonsense!

I'm sure a lot of people would agree with that! Supposing things turned out as the optimists believe, that in a few years time these superstring ideas prove to be along the right lines and that the difficulties you've referred to earlier are resolved, what sort of state would theoretical physics be in? We would have a theory which, on the face of it, would explain everything that happens in the universe. But do you in fact believe that? Do you think that a theory that identified the fundamental elements of the world would, in principle, solve everything – for example the problem of the origin of life and the origin of consciousness?

That's very big. You said a lot of things and I have to come back to them. Let's first start with the physics problem. It is certainly perfectly possible that some day, and perhaps with the superstrings, we'll get a theory which will explain all of our observations. And that, with a good mathematical analysis of the theoretical assumptions, we will show that the ratio of the masses of the muon to the electron is exactly as observed and all this other stuff works out. That the theory predicts correctly all of the aspects of nature and, perhaps, the theory contains within itself the best description of the origin of the universe. Okay? Now those problems are all part of fundamental theory. In the real world there are waves crashing against the shore, there are storms, lightning, wind, noise and so on, which we cannot analyse directly even if we did know all the laws of physics. In fact today we know enough of the laws of physics that, in principle, we should be able to analyse the waves crashing against the shore and the lightning and everything else. But the details of the interaction of winds and water and so on, is complicated – it's hard for us to analyse precisely.

Is it just complexity though, or are there any fundamentally new features that can arise?

Apparently to understand all those kinds of phenomena requires none of this advanced physics that we were previously talking about. The laws of quantum mechanics and of atoms and so on, without the nuclei being involved, are enough to explain the weather – although we really cannot explain the weather because of its complexity. I often use the analogy of a chess game: one can learn all the rules of chess, but one doesn't know how to play well. One can learn all the rules of physics; in fact we know them with sufficient accuracy in the realm of normal phenomena on Earth. Under ordinary circumstances we know them well enough. But that does not mean that we can analyse everything. In fact, natural phenomena are so elaborately complex that we cannot analyse them very well. Now I believe that the origin of life is one of those complicated phenomena. Physics has helped in understanding what molecules can do. That advance we've already made. What we're trying to do with the fundamental laws has more to do with the history of the universe and the final denouement of understanding every fundamental rule. The present situation in physics is as if we know chess, but we don't know one or two rules. But in this part of the board where things are operating, those one or two rules are not operating much and we can get along pretty well without understanding those rules. That's the way it is, I would say, regarding the phenomena of life, consciousness and so forth. Which way these issues work out and how philosophically they're going to work out is a nice question, but it's not a question that's waiting for physicists to obtain a complete grasp of the fundamental laws. We know the laws that govern the atoms that make life under certain conditions on the surface of the Earth.

But there are, of course, some people who maintain that when you examine systems that are sufficiently complex, new principles emerge – principles which are themselves maybe fairly simple in the way they operate, but which are not even in principle contained in the fundamental underlying physics.

Everything is right except your last statement. People may say that, but I don't see any reason to believe that. It's certainly true that when things become complex we use new principles to help us analyse things. For example, in chess, bringing the pieces toward the centre of the board increases their general strength. This is a principle which is not contained explicitly in reading the rules of chess, but which can be understood in terms of the rules of chess in an indirect fashion. The principle is obviously a consequence of only those rules and of nothing else. Yes, indeed, there are wonderful principles, ideas of valence, sound, pressure, and many other organising principles, which help to understand a complex situation. But to add that they're not contained in the fundamental laws is a misunderstanding. The fundamental laws have everything in them. It's just a question of finding convenient methods for analysing complex systems.

Yes, I didn't mean to suggest that these new principles might be incompatible with the underlying laws, only that the underlying laws are inadequate to encapsulate those principles.

I don't know what that means.

Well, that the principles might, for example, involve details about constraints, about the actual states of systems, which would not be present in the underlying laws themselves.

I don't know. I don't think so. There are many analogies you can make, such as in the analysis of computers. You find that if you have certain kinds of elements such as NAND gates, you can put them together to make any computer. But, on top of that, it's a great idea to have such concepts as the central processor and the memory section and so forth in order to understand the computer. Although it's still true that all these parts can be made from NAND gates, it's very useful to have those overlying principles. With things like wind, instead of worrying about the exact motion of each molecule which is how your laws are stated, it's more useful to know when large numbers of molecules are all moving in roughly the same direction. We can represent that by an average speed and so on and get an idea of wind, which isn't in the laws in an explicit fashion. The word 'wind' is not in the fundamental laws, but the fundamental laws contain the concept of wind. That's the way I think it happens.

Well, the issue I had in mind was the relationship between physics and cosmology. Although we may understand pretty well how the universe expanded in the big bang, the underlying physical laws don't seem to account for how the universe started. One has to put in special initial conditions. Do you think that we can understand the universe as a totality using physics, or are there additional principles that we need here?

That is a very interesting question because physics, so far, has always the following characteristic: a set of laws is given in such a form that if you tell the circumstances you can tell what'll happen next. In other words, if you put three atoms of this kind here, and five atoms of that kind there, one can tell what happens next. Those laws in fact have the property that they don't depend on the absolute time. The laws are the same then as they are now. So far, physics has never had in it an historical question of how did the laws get that way? There's no development. The laws of Newton, including the inverse square law of gravity, for example, don't say anything about *when* you're supposed to make measurements or *how* the laws came to be in time. So it was with the electromagnetic laws, quantum mechanics and so on. They are, I would say, local in time; you can apply the laws at any time. Therefore, they cannot apply in cosmology, because cosmology has to add something: how did things start? Only then can you work it out.

Now it's possible that those kinds of laws in physics may be incomplete. It might be that the laws change absolutely with time; that gravity for instance varies with the time and that this inverse square law has a strength which depends on how long it is since the beginning of time. In other words, it's possible that in the future we'll have more understanding of everything and physics may be completed by some kind of statement of how things started which are external to the laws of physics.

So you wouldn't agree with John Wheeler's idea that the laws of physics are capable of bringing the universe into being? You think we need something over and above those laws?

You have to be careful about what John Wheeler is saying, because by that I wouldn't know whether he means the laws of physics *should* or the laws of physics *do*. At the present time, the laws of physics do *not*. Even Wheeler will, I'm sure, agree that the present known laws of physics do not tell us how things began – they can't because of the way they're written. I know Wheeler and what he's probably saying is that the laws of physics, when they're fully understood, *will*. Well that's quite possible. That's what I was saying also, that maybe the future laws of physics in a completed form will not be of the kind that can operate at any time, but will describe the entire history of the universe without the need for any external proposition – as to how things started. But that's not the case at the present time.

How do you think of the laws of physics then? Do you think of them in some sort of Platonic way as existing independently of the universe, that is having an abstract existence in their own right?

Are you talking about now, or are you talking about the future?

Either.

Let's say now, alright?

Alright.

The problem of existence is a very interesting and difficult one. If you do mathematics, which is simply working out the consequences of assumptions, you'll discover for instance – of course, this is a minor proposition – a curious thing if you add the cubes of the integers. One cubed is one, two cubed is two times two times two, that's eight, and three cubed is three times three times three, that's twenty-seven. If you add the cubes of these, one plus eight plus twenty-seven and so on, and stop somewhere – let's stop here – that would be thirty-six. And that's the square of another number, six, and that number is the sum of those same integers one plus two plus three. We can try another number like five. One plus two plus three plus four plus five and then you square that; you'd get the same answer as if you'd cubed one, two, three and so on, up to five and added them. Alright? Now that fact, which I've just told you about, might not have been known to you before. You might say where is it, what is it, where is it located, what kind of reality does it have? And yet you came upon it. When you discover these things, you get the feeling that they were true before you found them. So you get the idea that somehow they existed somewhere, but there's nowhere for such things. It's just a feeling. This is human, we're psychologically struggling to understand. We find all these wonderful things, Bessel functions and their inter-relations, Fourier transforms, for example, they're really all there, and we just came upon them.

Well, in the case of physics we have double trouble. We come upon these mathematical inter-relationships but they apply to the universe, so the problem of where they are is doubly confusing. In the case of mathematics there's little doubt that these Bessel functions and so forth aren't anywhere, they had to be discovered, but somehow those relations existed before we discovered them. In the case of physics, because the laws are applied to the physical world and work, it gets even harder to say where they are. But they may be closer to reality than mathematical laws. Those are philosophical questions that I don't know how to answer. You can do a lot of physics without having to answer all that stuff. But it's fun to think about them.

Of course, at one time people used to believe that God explained the universe. It seems now that these laws of physics are almost playing the role of God – that they're omnipotent and omniscient.

On the contrary. God was always invented to explain mystery. God is always invented to explain those things that you do not understand. Now when you finally discover how something works, you get some laws which you're taking away from God; you don't need him anymore. But you need him for the other mysteries. So therefore you leave him to create the universe because we haven't figured that out yet; you need him for understanding those things which you don't believe the laws will explain, such as consciousness, or why you only live to a certain length of time – life and death – stuff like that. God is always associated with those things that you do not understand. Therefore I don't think that the laws can be considered to be like God because they have been figured out.

But they do seem to be all powerful and transcend the physical universe.

No. The physical universe obeys them. I don't know what you mean by transcend.

Well, if, as you were saying, the coming into being of the physical universe can be explained in terms of these laws, the laws must have existed in some sense before this universe began.

But we haven't got these laws yet. Are you talking about this hypothetical situation in which the physical laws describe how things started?

Yes.

Well, when we get there, I'll discuss the philosophy of the inter-connection with you. I can't answer without seeing them.

But do you believe that there are such laws?

I've no idea.

Well, do you think we're working towards some set of laws which are out there and to which our present theories are only an approximation?

Oh yes, of course. I get the feeling that I'm discovering laws that are *out there*, analogous to the feeling a mathematician gets when he discovers laws that he thinks are *out there*. But he knows that there's no physical place where his laws are. I know that my laws are useful to predict how the universe behaves, but again, I'm not very sure about where they are. It's a question I don't have to answer. I could do physics just as successfully without answering it. It doesn't mean I don't think about it, because you see I *have*: I've thought of the analogy. I find it very entertaining, delightful and amusing, but not profoundly important.

10

Steven Weinberg

Steven Weinberg is Professor in the Department of Physics at the University of Texas at Austin. His work has ranged across much of particle physics, quantum field theory, gravitation and cosmology, and he has made important contributions to each. He was awarded the Nobel prize for his work on unifying the weak and electromagnetic forces. Weinberg is an enthusiastic and eloquent proponent of superstrings, a subject on which he is now actively engaged.

About 100 years ago it was widely believed that physics was coming to an end, that is, a complete theory of the universe was just around the corner. The lessons of what we might call the new physics show that there is still a long way to go, yet there does seem to be a feeling that we are once again approaching the possibility of a Theory of Everything. Do you think this is another false hope, or is there really some chance that we are now moving to the stage where we will be able to write down just a single expression or principle that will govern all of nature?

I think we physicists have learned to be appropriately modest. We have a goal of a unified view of nature in which, at least in some attenuated sense of principle, everything will follow from a few simple underlying laws (although that will never really help us to understand trees and people). But we know what the difficulties are. We know that it's very difficult, for example, to bring something as different as gravitation into the same picture of forces as the nuclear forces and the electromagnetic force. There has been a great deal of progress in the last decade or so in achieving a unified view of the forces that act on elementary particles at accessible scales of energy, excluding gravitation, but it's very difficult to make that final step and bring gravity into the picture.

What are some of the latest ideas for bringing gravity into the picture?

Well, if you had asked me that a few years ago, I would have said there aren't any. In fact there is an idea which has been around I suppose since 1974, called string theory, or in its later version, superstring theory. It was originally invented around 1968 in order to try to understand the strong nuclear forces, the forces that act within the nucleus of the atom to hold it together. It proved to be a terrible failure in that role. One of the troubles was that this theory predicted particles of a type with zero mass that would clearly not fit into anything we knew about the structure of the nucleus. In 1974 it was suggested by John Schwarz and Jöel Scherk that these theories needed to be reinterpreted, not as the theory of nuclear force, but as a unified theory of *all* forces, including gravitation, and that the particles of zero mass that appeared in these theories, so embarrassingly when they were intended to be theories of nuclear force, were in fact to be identified with the quanta of gravitational radiation, the so-called gravitons.

How have these theories developed in the intervening years?

For years after 1974 very little attention was paid to these string theories. I can speak for myself and say that I paid them no attention whatever. We particle physicists were having a great deal of fun in developing the now standard view of elementary particle physics, the unified picture of weak and electromagnetic – and we hope also strong – interactions, which was so very successful and so very well confirmed by a series of brilliant experiments. We tended to put gravitation off for the distant future because we saw no hope of bringing it into the picture. Those who had participated in the early days of developing the so-called string theories, continued to work on them, pretty much ignored by the rest of us. It's only in the last few years that their work has re-emerged in the general consciousness of particle physicists, partly because we have been so frustrated with all other efforts, and partly because of some stunning mathematical developments.

These theories have been found to be consistent mathematically in a way that had not been thought to be possible. The consistency, moreover, apparently only obtains for a very limited class of such theories, so that these string theories have a quality that physicists are always looking for – inflexibility. String theories are very rigid. There is not an infinite variety of possible games we can play, as there are for most of the other kinds of theories we have been speculating about for the last decade. Perhaps there is only one, or a very small number of games, and by playing these games we think therefore we might have a chance of really making progress.

Now these strings actually inhabit a ten-dimensional universe in the preferred version of this theory, isn't that correct?

Yes, more or less. That was, of course, one of the main reasons why string theory did not immediately catch on. It is very beautiful mathematically, everything fits together in quite a wonderful way, but only if the theory is formulated originally, in fact, in twenty-six dimensions, and then with further modificiations to make it appear more realistic, ten dimensions. That is nine space and one time dimensions. Now that's not observed of course. Of all the numbers that have been measured experimentally, clearly the one that we are most confident of is the number four for the number of our spacetime dimensions. So when these ideas were first mooted in 1974, they attracted very little attention simply because they seemed to be immediately out of the question. We could not imagine ourselves being happy with a theory of gravitation in ten dimensions. What we wanted was one in four spacetime dimensions. Well, one of the things that has happened in the last decade is that physicists have rediscovered the old idea, which goes back to Theodor Kaluza in 1921, that we may very well live in a higher-dimensional universe in which all but the familiar four dimensions of spacetime have become tightly curled up to a circumference so small that we don't normally observe them. This idea was originally introduced by Kaluza with a certain amount of encouragement from Einstein in order to try to explain the other forces of nature, like electromagnetism, in terms of pure gravity acting in a higher dimensional space. And for that reason also, the idea was revived in the early 1980s and has been a subject of active effort on the part of theoretical physicists in the last few years. I think it was the revival of the old ideas of Kaluza and Klein and others, that have prepared the ground, and dissolved our scepticism for string theories formulated in ten dimensions.

The picture is that although the theory in some sense is fundamentally a ten-dimensional theory, six of the dimensions are lost to our view through what is called 'spontaneous compactification'. Dynamical effects cause them to curl up so tightly that we are simply not aware of their existence. So we go about blithely assuming that there are only three space and one time dimensions. But there may very well have been a time in the early stages of the universe when there would have been manifest to any scientist who might have been around then (not that there were any!) the full nine space dimensions plus one time dimension.

How is it that a string can appear to us at low energies to be like a particle? At first sight you would think that these are very different types of entities.

The string (imagine for, example, a closed string), will exist in an infinite number of modes of vibration. Each mode of vibration appears as a single species of particle. As you go up to higher and higher modes of vibration, there are more and more wavelengths fitting around the loop of the string, and you are dealing with particles of higher and higher mass. The lowest modes of the string – the particles of very small mass – are the ones we see in our laboratories, the ordinary particles. The other ones – the higher modes – are probably going to be unobservable for the foreseeable future.

Is it correct to think of the string at low energies moving like a rigid body and therefore appearing to be particle-like, but as the energies are raised so the string begins to wiggle about and therefore behave differently?

Very roughly, yes. That's a rough description. If we consider a collision of two particles of the type we are familiar with, that would correspond to low modes of vibration of the string, and we calculate the forces that would be produced by the exchange of gravitons, we might come to the conclusion that these forces were infinite, because that is what we had always found when we tried to deal with gravitation on the quantum level. However, although the particles that we deal with in the laboratory correspond to low modes of vibration of the string, when they are exchanged in producing forces they are exchanged in all modes, and this sum of an infinite number of modes adds up to give a finite result, quite wonderfully – almost miraculously – and yet quite believably, mathematically.

How should one envisage the superstrings in terms of describing the difference between an electron and a neutrino, because in the case of an electron we have a charged particle, and in the case of a neutrino we have an uncharged particle? What would be the difference in the superstring description of these two particles?

In a way, that's not an appropriate question. The description of the particles with which we are familiar, electrons, neutrinos, protons and so on, at the energies at which we normally study them, is not going to change. It's presumably going to remain the theory that has become known as the standard model. In that theory the electron and the neutrino are two different members of a family of particles, and it's true that the electron has a charge, which means that it interacts directly with the electromagnetic field, and the neutrino doesn't. But the neutrino has other interactions with members of a family of interactions, of which electromagnetism is just one member. It all works out and is perfectly nice and symmetrical and pretty, except for the fact that the symmetries that relate the electron and the neutrino, and the electromagnetic forces and other forces, are all broken.

All that is old stuff, and is not likely to be changed by the advent of superstring theory. The question is, can superstring theory reveal the standard model as a low energy approximation? In the low energy approximation, the strings are seen as point particles. Depending on the particular mode of vibration, they are electrons or neutrinos or whatever, and we have to try to understand whether the standard model, with all of its particles (including electrons and neutrinos) comes out of the superstring theory. That's the main question.

More detailed questions, like why the electron has a charge and the neutrino hasn't, are answered the same way as in the standard model, because the point of superstring theory is not to supplant the standard model, but to rederive it from the low energy approximations, with all its loose ends tied up.

Unfortunately, theorists have not yet succeeded in deriving the standard model as a low energy limit of the superstring theory. They have come tantalizingly close; superstring theories naturally have ingredients in their low energy limits which look very much like the standard model but so far no-one has succeeded in making it come out exactly right.

I find your question an awkward one. It's like asking 'how in general relativity do you work out the shape of a suspension bridge?'. We work out the shape of a suspension bridge using Newtonian gravity, and one of the nice things about general relativity is that it has Newtonian gravity as an approximate version, which is relevant on the scale of distances characteristic of the Earth's surface. So I wouldn't reconsider the shape of a suspension bridge after the discovery of general relativity, and I wouldn't reconsider all the previous successes of the standard model after the development of superstring theory.

Do the superstrings have charge spread out all over them or is it localized?

There are different ways that it can arise. The original answer would have been that electric charge can only arise in an open string theory, and that electric charge, like the other attributes having to do with weak interactions as well as electromagnetic interactions, are labels attached to the ends of the string. You can think of the string as carrying these quantities around at its end.

That was the original picture. There are much more subtle pictures now in which charges have something also to do with the way the ten dimensions reduce to four dimensions.

Another question related to this, is how one gets a multiplicity of different types of particles – quarks, electrons, neutrinos and so forth.

The strings are vibrating in all these extra dimensions and that leads to a lot of different modes. It is the extra dimensions (or other extra physical variables) which produce the many different modes. In fact, that's one of the encouraging things about string theory. Because of that it's natural to find multiple generations of particles, not just the lowest generation with the light quarks and the electrons, but also the next generation which includes the strange quarks and the muons and so on. In fact, most of these models have too many generations. In one of the earliest papers in which one tried to get a specific low energy theory out of the string theory, they had something like one hundred generations. It has to do with the fact that the string can vibrate in these other dimensions.

Am I correct that all observed particles correspond to just the lowest frequency mode of vibration?

Yes. These are the lowest modes. The next mode would be hopelessly too heavy to have ever been seen. There are lots of those lowest modes, all of which correspond to particles which from the point of view of the scale of energies typical of string theory, are essentially massless.

The natural scale of mass built into the string theory is about 10^{19} GeV and all of the particles that we are talking about are essentially the zero mass modes of the string.

But the vibrations within the extra dimensions don't involve the Planck mass?

Right. There is an infinite number that do, and a finite but still relatively large number that don't, and those are the ones that we see as massless particles. Actually not *quite* massless because there are more subtle effects which can give them a little mass, but still a very tiny mass compared to the Planck scale, 10^{19}

GeV. There may be only a few hundred of those light particles, whereas there is an infinite number which have these very high masses extending up and up.

Some people say that one has to answer that problem in terms of 496 different types of charges.

That's the celebrated Green–Schwarz theory, which is by now a rather old-fashioned theory, in which you have an open string with loose ends. The charges are carried at the ends, and the number of ways you can combine them adds up to 496. It turns out that that is the only possible number that would lead to a quantum mechanically consistent theory constructed along these lines.

What has happened since the ground-breaking work of Green and Schwarz, is that a number of other ways of constructing satisfactory theories have been discovered which can lead to other numbers of charges. It may not be 496, but we think there is some fairly large yet finite number of possible values of different charges.

How should one envisage these charges?

As my friend Abdus Salam is fond of saying, we should expect nature to be simple in its principles, but not necessarily simple in its structures. The string theory starts with an extremely simple set of assumptions about the way the world is. Then there is some rather complicated mathematics, which leads to a fairly complicated picture of what one would see experimentally at the energies that are accessible to human beings. I don't think we should be disappointed that the universe is complicated, or that the string theory predicts it will be complicated. The point is not how complicated the output is, but how complicated the input is. The input, the fundamental assumptions, are extremely simple, and I think it is that more than anything that makes these theories so exciting and so beautiful. One is not whipping up a very complicated soup just by throwing in ingredients and tasting it each time to see how you like it. The recipe is fixed from the beginning and is extremely simple, even though the final menu turns out to be fairly elaborate. The 496, for example, may seem like a large elaborate structure, but within the framework that Green and Schwarz were pursuing it's the only possibility, and it just automatically comes out of very simple assumptions. I would always look at the simplicity of the assumptions, not the simplicity of the consequences.

Is superstring theory able to say anything specific about proton decay?

Yes. In fact, some of the superstring theories in their low energy limits ran into trouble because they gave rather alarmingly high rates for proton decay. That's one of the things that is going to be a very important constraint on whether we call a superstring theory a success or not. If it predicts that protons decay with a lifetime like that of the pi meson, then clearly those solutions are going to have to be thrown out. Proton decay is something that's going to be looked at very closely by anyone who tries to find a low energy version of a particular superstring theory, but I don't think there is anything you can say that's generic to *all* of superstring theory. I don't think you can say that superstring theory makes proton decay inevitable or makes it impossible, or makes it too fast or too slow. It's just a detail that will come out of a particular solution.

In the early days of string theory, Green and Schwarz demonstrated the anomaly cancellations in the theory. Is it correct to say that at the time it looked like the anomaly cancellations uniquely selected a theory with ten or twenty-six dimensions?

Yes, it is correct. One studies a lot of different possibilities but we are not sure whether these possibilities are different theories or whether they are different solutions of a single theory. If they are different solutions of a single theory, we don't yet know what are the physical principles that determine which solution corresponds to the world we see. Certainly there are more general possibilities known now than were known at the time of Green and Schwarz's work, but we're not sure really how general they are, and it may turn out that they all boil down to the same thing.

It's a very bad time to make statements about the ultimate meaning of string theory because there is so much uncertainty right now about which of these many different kinds of solutions are really solutions, and which are independent theories.

Could we clarify the point about the 496 charges? What sort of charges are these?

In the Green–Schwarz picture those charges are put in essentially by hand. They don't have to do with the extra dimensions, they are just charges that are carried at the ends of the open string, and that number is needed in order to balance other effects that would produce anomalies and would therefore destroy the consistency of the theory. But they are put in by hand. You just say 'let the theory have this number of degrees of freedom'. Now when I say the charges are put in by hand, I mean that you consider a multiplicity of strings carrying all the different charges, but you then find automatically that the forces that act on those charges – the electromagnetic force, the weak force and all the other forces that are produced by these charges, and act on these charges – all those come out automatically. But the charges themselves are put in in a somewhat *ad hoc* way as required by the condition of mathematical consistency.

It seems that the picture has changed slightly now, because people are beginning to see other ways of formulating the theory. Most recently, some people are proposing to formulate it in four dimensions.

Yes, I was going to correct my own remarks in just that way. I talked about the extra six dimensions wrapping themselves up, but that's not necessarily the way that one thinks about it now. One thinks about the theory as formulated in four dimensions but with some extra variables, which can, in some cases, be interpreted as coordinates of extra dimensions, but needn't be. In fact, in some cases they *cannot* be. So you really understand what the general possibilities are if you give up this geometrical picture of the extra degrees of freedom and just think of four good old spacetime coordinates plus a lot of extra variables which are needed in order to make the theory consistent.

There is a consistency condition which *requires* these extra variables and, in fact, specifies the menu of those extra variables, although we don't know exactly what the rules are for determining that menu. But they're not things that we can add freely as the spirit moves us. We have to add extra degrees of freedom beyond the four coordinates of spacetime. The various ways of doing it are tightly circumscribed by the conditions of mathematical consistency. We don't yet know exactly how to find a general way of satisfying those conditions, or how to prove that we have in any case found a completely satisfactory solution. But that's the direction of current research. The original picture of ten dimensions, of which six get curled up, is just seen as one special case.

If one is no longer proposing that these extra variables correspond to higher dimensions, is it possible to give them any physical interpretation?

I think not. The final theory is going to be what it is because it's mathematically consistent. Then the physical interpretation will come only when you solve the theory and see what it predicts for physics at accessible energies. This is physics in a realm which is not directly accessible to experiment, and the guiding principle can't be physical intuition because we don't have any intuition for dealing with that scale. The theory has to be conditioned by mathematical consistency. We hope this will lead to a theory with solutions that look like the real world at accessible energies.

I'm afraid that the usual physical insight based on experience with experiments in physics isn't a great deal of help here.

Mike Green claims that we will have to change our concept of space and time to make it string-like. At the moment string theory is formulated on a classical background.

I think that in these theories space and time may not turn out to have overwhelming importance. Space and time coordinates are just four out of the many degrees of freedom that have to be put together to make a consistent theory, and it's only we human beings who give them that peculiar geometric significance which is so important to us. I think in that respect I'm not representative of most string theorists. Most string theorists are trying to find something beautiful and geometrical underlying string theory – similar to the principles that Einstein found in general relativity. They may succeed. But I suspect that that may be a misleading analogy and that what we are going to have is not so much a new view of space and time, but a de-emphasis of space and time. The spacetime coordinates may turn out to be just four out of the ten – or fifteen or twenty-six or whatever it is – degrees of freedom that are needed to describe the theory. Their geometric significance will arrive *after* the fact, rather than something that appears in the fundamental principle.

There was a lot of fuss made a few years ago because the theory looked as though it was definitely going to be finite. To date, as I understand it, the theory has only actually been proved to be finite in an approximation expansion to a certain number of orders. But wasn't this also the case with the old supergravity theory, which was also hailed to be a finite theory, but in the end wasn't?

I think there's a difference. With supergravity, the arguments which showed that it was finite were clearly arguments that were only relevant to the lowest orders of perturbation theory. The arguments took the form of showing that the infinities that *could* arise, *would not* arise in the first or the second approximation of perturbation theory.

The arguments in the string theory are very different. There are hand-waving arguments that are not at all rigorous, perhaps not entirely convincing, that the theory *ought* to be finite to all orders. Then when one works out how it really looks in the lowest orders of perturbation theory, one finds that those hand-waving arguments really work. In other words, the reasons why the low orders of perturbation theory were expected to be finite in supergravity were specifically limited to low orders of perturbation theory, whereas the reasons here are very general and are borne out by low orders of perturbation theory. So I think the situation is quite different.

With supergravity I would have said that one would have to be an extreme optimist to have expected that theory to be finite beyond the first few orders of perturbation theory, where one *knew* it was finite for very special reasons. With superstring theory I think finiteness is a reasonable guess. I'd be more surprised than not if it weren't finite.

What is your response to some of the criticisms that have been raised against superstring theory?

I think one does what one can – that's the first principle in physics. You do whatever you can to make progress. I would be delighted if there was the same kind of happy collaboration between theory and experiment now that there was fifteen years ago, when theorists were trying out new ideas and experimentalists were testing them, and the experimentalists were discovering new phenomena, and the theorists were responding. Unfortunately, we had so much success in those days, that that chapter has come to an end. It may reopen with the next generation of accelerators. We hope that when the SSC and LEP, and perhaps even the existing accelerators like the Tevatron collider, are brought into operation that we'll have that kind of give-and-take again. That would be wonderful. But in the meantime, one has to do what one can.

One thing you might do is to try to be very clever and think of ways of making progress beyond what we already know, using existing accelerators and other facilities. I'm awfully glad that there are people who are doing that, and I hope they succeed. It hasn't been very fruitful so far. We have not made any real progress beyond the standard model.

The other possibility which one should also try is to make a big leap, and go to the most fundamental level, and try to understand what's happening in a deductive way from some very simple elegant principles. It's worth doing if you have some good ideas. It seems to me that superstring theory is an awfully good idea and it's worth pursuing. I don't think everyone should work on superstring theory, and I don't think that everyone should work on phenomenology and low energy physics. I think people ought to do what they can. I do think however that superstring theory is giving the education of our graduate students a strong mathematical flavour. It's good that they're learning all this mathematics but I'm worried that some of them don't know what a pi-meson is. In fact, I'm teaching a course this year in Austin called 'Elementary particle physics' in which I start with the discovery of the electron by J.J. Thomson in 1897, and go through all the agonizing experiments and theorizing that took ninety years to bring us to the present understanding.

So I'm very sympathetic to immersing oneself in the phenomena and trying to grapple with them theoretically, but I think it's also worth trying to look ahead, jump over seventeen orders of magnitude in energy and look up to the Planck scale where the final answer may lie. Whether string theory will have been a good idea depends on what comes out of it. But it's crazy not to try because it is very beautiful, very promising and it's had qualitative successes so far in making a lot of things come out right when it wasn't clear how they could ever come out right – things having to do with gravity. It's worth a try.

It's very difficult to see what the consequences of these string theories are, to know what predictions the experimentalists or the astronomers should set about trying to verify or refute. We really don't know yet what these theories actually predict, in the way of new forces, or particles that may be left over from the early universe, or may be found in our accelerators. There are hints that these theories involve the presence of new forces acting at the ordinary scale of elementary particle physics in addition to the ordinary strong, weak and electromagnetic forces. But the details haven't been fleshed out.

An increasing fraction of theoretical physicists, especially the younger ones, are working in this area, but the mathematical obstacles are very formidable and it may very well turn out, as has often happened before, that although these theories will present, at first, a resemblance to the real world, that there'll be insuperable obstacles to finding an interpretation that corresponds to physical reality. But it's certainly going to be a lot of fun in the next few years to work that out.

GLOSSARY

Aether. A hypothetical medium supposed to fill all space, in which electromagnetic waves propagate. The idea was discredited by the theory of relativity.

Atomism. The theory, originated by the Greek philosophers Democritus and Leucippus in the fifth century BC, that all matter is composed of indestructible microscopic particles.

Baryons. Heavy hadrons that consist of three quarks.

Bosons. Name given to the class of particles with intrinsic spin either zero or an even multiple of the fundamental unit of spin.

Chirality. The technical term for a system or object which possesses a definite left- or right-handedness.

Fermions. Name given to the class of particles with intrinsic spin equal to an odd multiple of the fundamental unit of spin.

Feynman diagrams. A technique for studying particle interactions in terms of diagrams. Although the diagrams are physically suggestive, they are really only schematic, and represent terms in a calculation rather than real processes.

Gauge theory. A theory in which a force is described in terms of a field that possesses certain abstract symmetry properties.

Gluons. Particles or quanta that convey the strong force between quarks.

Grand Unified Theories (GUTs). Theories which attempt to provide a unified description of three of nature's four basic forces - the electromagnetic, weak and strong forces.

Gravitinos. Hypothetical particles that are partly responsible for conveying gravitational forces according to the supersymmetric theory of gravity.

Gravitons. The particles or quanta of the gravitational field whose exchange between particles of matter can be regarded as responsible for gravitational forces.

Hadrons. The collective name given to subnuclear particles (generally heavy) that experience the strong nuclear force.

Leptons. The collective name given to (generally light) particles of matter that experience the weak, but not the strong, nuclear force.

Mesons. Intermediate mass hadrons that consist of a quark bound to an antiquark.

Muons. Members of the class of particles known as leptons. They are essentially identical to electrons, except for being much heavier, and unstable.

Neutrinos. Leptons with zero electric charge and probably zero mass. They are so weakly interacting that they are almost impossible to detect.

Parity. Name given to the mirror reflection properties of subatomic particles.

Phenomenology. Literally, the study of phenomena. The word is used colloquially to mean the pragmatic analysis of experimental data without too much concern for underlying theory.

Photons. Particles or quanta of light and other electromagnetic waves. Photons can be regarded as responsible for conveying the electromagnetic force.

Positrons. Antimatter partners of electrons. A positron is an electron with all qualities except mass reversed. In particular it possesses a positive electric charge; hence the name.

Planck's constant. Originally this constant, denoted by h, was introduced by Max Planck as the ratio of energy to frequency of photons, which is a fixed and universal number. Planck's constant permeates quantum theory, and reappears (usually divided by 2π) in many other contexts, e.g. as the fundamental unit of intrinsic spin.

Principle of equivalence. Name given by Einstein to the equivalence of acceleration and gravitation. The principle is demonstrated in its most familiar form by the observation that all bodies fall equally fast under gravity.

Quarks. Elementary constituents of hadrons (nuclear particles). Quarks combine in threes to make baryons (e.g. protons) and in pairs to make mesons.

Reductionism. The philosophy that all physical processes and systems can ultimately be understood solely in terms of their most primitive constituents.

Strong force. The force that acts between hadrons (nuclear and associated particles). In its modern form, the strong force is regarded as originating with the force between quarks.

Supergravity. Theory in which gravity is treated as part of a supersymmetric description of spacetime geometry.

Supersymmetry. An abstract geometrical symmetry that unites bosons and fermions in a common description. Supersymmetry forms the basis of most modern attempts at providing a quantum theory of gravity, and is an essential ingredient in the superstring theory.

Topology. The branch of mathematics that deals with the way that lines, curves, surfaces, etc. connect to themselves. Topology is little concerned with geometry as such (that is, with sizes and shapes) but concentrates on such issues as how many knots there may be in a line, or how many holes in a surface.

W and Z particles. These convey the weak force. They were discovered in 1983, although their existence had for some time been predicted on theoretical grounds.

Weak force. One of the four fundamental forces of nature, the weak force acts on all particles of matter, though it is often swamped by the much stronger strong and electromagnetic forces. The most conspicuous effect of the weak force is to cause beta radioactivity in nuclei.

INDEX